CHICAGO PUBLIC LIBRARY
R01162 36238

D1566540

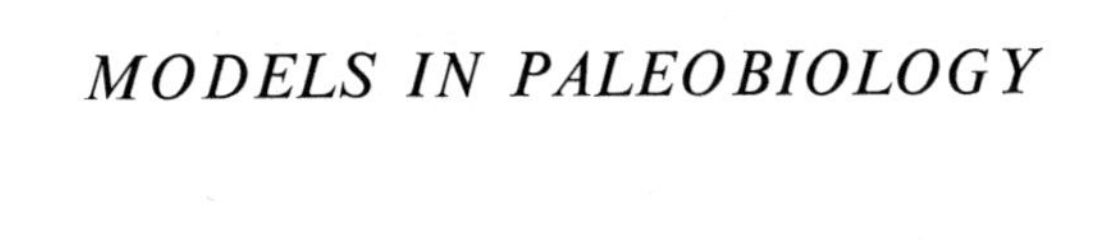
MODELS IN PALEOBIOLOGY

THE CONTRIBUTORS

J. Robert Dodd, Department of Geology
Indiana University, Bloomington, Indiana 47401

Robert G. Douglas, Department of Geology
Case Western Reserve University, Cleveland, Ohio 44106

Niles Eldredge, The American Museum of Natural History
Central Park West at 79th Street, New York, New York 10024

Michael T. Ghiselin, University of California, Berkeley
Bodega Marine Laboratory, P.O. Box 247, Bodega Bay, California 94923

Stephen Jay Gould, Museum of Comparative Zoology
Harvard University, Cambridge, Massachusetts 02138

Anthony Hallam, Department of Geology and Mineralogy
University of Oxford, Oxford, OX1 3PR, England

Ralph Gordon Johnson, Department of the Geophysical Sciences
University of Chicago, Chicago, Illinois 60637

Ismail A. Kafescioglu, Department of Geology
The American University of Beirut, Beirut, Lebanon

David M. Raup, Department of Geological Sciences
University of Rochester, Rochester, New York 14627

Thomas J. M. Schopf, Department of the Geophysical Sciences
University of Chicago, Chicago, Illinois 60637

Daniel Simberloff, Department of Biological Science
The Florida State University, Tallahassee, Florida 32306

Francis G. Stehli, Department of Geology
Case Western Reserve University, Cleveland, Ohio 44106

James W. Valentine, Department of Geology
University of California at Davis, Davis, California 95616

MODELS IN PALEOBIOLOGY

Edited by Thomas J. M. Schopf

FREEMAN, COOPER & COMPANY
1736 Stockton Street, San Francisco, California 94133

Printed in the United States of America

Library of Congress Catalogue Card Number: 72–78387

SBN 87735–325–5

CONTENTS

Part IV DISTRIBUTION

PART I

Introduction

ABOUT THIS BOOK

The motivation for this book can be traced to a chance conversation with a graduate student at the Atlantic City meeting of the Geological Society of America in 1969. In answer to my casual inquiry about what he was working on, he stated that his professor had made a collection of brachiopods from a certain formation, that these fossils had never been described, and that this description would make a good problem. The tacit assumption was that because the description had not previously been made, it was worth doing. These offhand remarks reminded me of similar conversations with other students in other places, and I began to wonder again why it is that this type of descriptive study is so popular. The answer, I decided, was that this particular student was unaware of the various alternative strategies of research available in invertebrate paleontology.

But what strategies are available to us? As John Platt (1964) wrote: "We praise the 'lifetime of study,' but in dozens of cases, in every field, what was needed was not a lifetime but rather a few short months or weeks of analytical inductive inference. In any new area we should try, like Roentgen, to see how fast we can pass from the general survey to analytical inductive inference. We should try, like Pasteur, to see whether we can reach strong inferences that encyclopedism could not discern." In a field where the description of faunas is held in such high esteem as it is in systematic paleontology, these may be uncomfortable words to some of us. But the issue as to what is the best research strategy does not go away simply because our traditions support one particular viewpoint. Platt may be right; what then!

This book resulted from a symposium. We remind our classical friends that according to Webster a symposium may be "a drinking together, with . . . conversation. . . . (a social gathering at which) there is a free interchange of ideas . . ., a particular subject is discussed (resulting in) a collection of opinions on a subject."

Hence this book. Implicit in it is the belief of the group that paleontology has collected much of its data and basic theses. Explicit is the belief that henceforth paleontologists may and should turn to broader horizons and interpretive themes. Thus we offer these models as examples, or rather as suggestions, for our fellow workers.

Prevailing ideas generally dictate the data that paleontologists traditionally collect. Such ideas are increasingly dependent on the Recent in order to establish models which can be tested in a paleontological context. We look with awe at the results which have followed the application of fundamental chemistry in a geochemical context, and the use of the Cartesian methodology in molecular biology. Our fossils are amenable to more than the traditional methods of analysis. We wanted to write a book which would take as its goal the self-conscious use of models in paleontological research.

The results are mixed in their degree of excitement, astuteness, and readability. Different chapters must of necessity take as their goal different aims, partly because of the various stages of development of different parts of paleontological research, and partly because of the style and judgment of each author. Original work versus review varies considerably from one chapter to the next. Were it not so, the book would not have been prepared by those interested in paleontological research, or for that matter by any group of practicing scientists. A brief statement by the Editor about each chapter and its intended role precedes it.

We should comment on the intended coverage. In the planning, we really have tried to go from the morphology of individuals on the one hand to the way in which ecosystems are put together on the other hand. Paleontologists have traditionally been concerned with three general areas: patterns and processes of morphology *per se*, of evolution, and of distributions of organisms. We have attempted to emphasize the philosophical rationale guiding the development of these same divisions by successively considering morphology (Raup; Dodd and Schopf), populations and speciation (Hallam, Eldredge, and Gould; Stehli, Douglas, and Kafescioglu; and Ghiselin), and distributions (Johnson; Simberloff; and Valentine). The mechanism for doing this is to discuss a field in terms of the models that are currently of use as conceptual guides for planning experiments and observations. Some authors focus attention on particular models which they feel are of outstanding value. Despite the variations in degree and method of treatment, the general goal remains the same: to direct attention toward the conceptual foundations of paleobiology.

We of course do not imply that some topics not prominently included (such as genetics, development, etc.) are without serious interest. Any judgment of our sample on the basis of what it might have been must take into account who had enough of a conviction to prepare a manuscript on it, in a prescribed length of time. In any event, we know of no closely comparable book in the field and we hope that the book will help to provide focus for the analytical approach in paleobiology.

The chief subject is marine invertebrate paleontology. Those who work on fossil vertebrates or fossil plants have less need for much of what is said because their biological background is so much better. Nevertheless, the methodology is worth emphasizing. We need one other disclaimer. The other major part of invertebrate paleontology—biostratigraphy—is much more closely related to theoretical concepts of historical geology. We hope that others may deal with that topic on another occasion.

Authors were approached in the spring of 1970, and manuscripts were submitted in March, 1971. The early planning received considerable support from J. W. Valentine and F. G. Stehli. Manuscripts were extensively reviewed, and were resubmitted in July, 1971. The papers were presented (in various degrees of fidelity to this written version) at the annual meeting of the Paleontological Society, and thc Geological Society of America, at Washington, D.C., November 2, 1971. During the course of preparation of the book, I have often benefited from the advice of R. G. Johnson and S. J. Gould.

Many of the authors commented on chapters of other authors, and so served as internal critics. We also greatly benefited from external critics, some of whom read more than one chapter at my request. These special critics, not all of whose advice was followed, include R. H. Benson, A. G. Fischer, W. A. Berggren, S. M. Bergström, F. T. Manheim, E. O. Wilson, L. Van Valen, J. Hopson, K. W. Kaufmann, R. Osman, J. Sans, G. A. Curtis, and J. Jacoby.

In addition, T. J. M. Schopf and A. Hallam thank the Geological Society of America for travel funds which permitted Hallam's attendance at the conference. F. G. Stehli, R. G. Douglas, and I. A. Kafescioglu are indebted to E. Mayr and K. Flessa for thoughtful criticism, helpful discussion, and/or the use of unpublished data and to Flora Rhodes Burkholder for figure 6–7; they also gratefully acknowledge the financial support of the National Science Foundation (Grant No. GA-16827), the Petroleum Research Fund of the American Chemical Society (Grant No. 3485-AZ), and the U.S. Atomic Energy Commission (Contract AT-11-D-1796). An informal discussion of the ideas presented in the chapter by R. G. Johnson was given at the Penrose Conference on Marine Ecology and Paleoecology held at Pacific Grove, California, in December, 1970. He greatly profited from the suggestions and misgivings of the participants. The original research referred to in Johnson's chapter was supported by the Federal Water Quality Administration (Grant #18050 DFP). D. S. Simberloff thanks Michael P. Johnson for several suggestions which improved the manuscript considerably. J. W. Valentine acknowledges a special debt of gratitude to Dr. John Warme, Rice University, for extensive discussions of modeling in living and fossil ecosystems from which some of the ideas presented in his chapter have come.

Finally, our first commitment is to a reinvigoration of paleontology. To achieve that, we would reach as wide an audience among working paleontologists and their students as possible, hoping thereby to light some new paths for inquiry. Thus contributors have forgone royalties—and reprints—and the publisher any profits. The result is a price modest enough to encourage purchase by the people we want most to reach—the working paleontologist and his students. Should support outrun our most optimistic hopes, any material profits will go to the Paleontological Society, which has supported in spirit our efforts.

THOMAS J. M. SCHOPF
Editor

Acknowledgments

We thank the following for permission to illustrate previously published material: Henry Stommel and the American Association for the Advancement of Science, for Figure 1, in Varieties of Oceanographic Experience, H. Stommel, *Science*, v. 139, pp. 572–576, 15 February 1963, Copyright 1963 by the American Association for the Advancement of Science; the Chemical Society of Japan, for Figure 4, in The behavior of various inorganic ions in the separation of calcium carbonate from a bicarbonate solution, Y. Kitano, *Bull. Chem. Soc. Japan*, v. 25, pp. 1973–1980, 1962; The University of Chicago Press, for Figure 8, in Biologic problems relating to the composition and diagenesis of sediments, H. Lowenstam, in *The Earth Sciences: Problems and Progress in Current Research*, T. W. Donnelly, Ed., The University of Chicago Press, pp. 137–195; Maxwell International Microforms Corporation, for Figure 5, in Fractionation of the stable isotopes of carbon and oxygen in marine calcareous organisms . . . The Echinoidea, J. N. Weber and D. M. Raup, *Geochimica Cosmochimica Acta*, v. 30, pp. 681–703, 1966; Cambridge University Press, for Figure 5, in The use of *Gryphaea* in the correlation of the Lower Lias, A. E. Trueman, *Geological Magazine*, v. 59, pp. 256–268, 1922; McGraw-Hill Book Company, for Figures 1–14 and 1–15, in *Invertebrate Fossils*, R. C. Moore, C. G. Lalicker, and A. G. Fischer, 1952; Stephen Jay Gould, for Figure 20, in An Evolutionary Microcosm: Pleistocene and Recent History of the Land Snail *P.* (*Poecilozonites*) in Bermuda, S. J. Gould, *Bull. Museum of Comparative Zoology*, v. 138, pp. 407–531, 1969; McGraw-Hill Book Company, for Figure 637, in *The Course of Evolution*, J. M. Weller, McGraw-Hill Book Company, 1969; Duke University Press, for Figure 2, in Experimental zoogeography of islands. A model for insular colonization, D. Simberloff, *Ecology*, v. 50, pp. 296–314, 1969; Princeton University Press, for Figures 44, 45, and 51, in *The Theory of Island Biogeography* (Copyright © 1967 by Princeton University Press), R. H. MacArthur and E. O. Wilson, Princeton University Press, 1967; Society for the Study of Evolution, for Figure 1, in Taxonomic diversity of island biotas, D. Simberloff, *Evolution*, v. 24, pp. 23–47, 1970; Biology Department, Brookhaven National Laboratory, for Figure 5, in The species equilibrium, E. O. Wilson, *Brookhaven Symposia in Biology*, v. 22, pp. 38–47, 1969; W. H. Freeman and Company, for quotation on pages 75–76 of *Worlds-Antiworlds: Antimatter and Cosmology*, H. Alfvén, Copyright © 1966.

This editor, for the freedom to think quietly about many of the issues, acknowledges his debt to the Marine Biological Laboratory, Woods Hole, Mass.; also his debts to the creative inspiration of James Morton Schopf, and to the encouragement of his mother, Esther, and his brother, James William; he dedicates to his wife his contributions.

1

VARIETIES OF PALEOBIOLOGIC EXPERIENCE

Thomas J. M. Schopf

Editorial introduction. The purpose of this first chapter is fourfold. First, Schopf presents a brief resumé of the historical emphasis in invertebrate paleontology followed by a mention of the spectrum of ways in which the word "model" is used. Secondly, he introduces one way to think about the notion of optimization. In the context of different fossil groups and different levels of biological organization, this is a key feature of a large amount of paleontological work. Various aspects of optimization recur in nearly every other chapter in this book.

Thirdly, he presents a way to think about the dimensions of the units used in paleobiological work. For example, a great difference exists in the length of time over which one considers a "community" to have changed; the time scale will vary according to the model (equilibrium or historical) of a community that one has in mind. Each is justified under different conditions. Lastly, an appendix is given with a brief review of the concepts guiding textbooks in invertebrate paleontology since Pictet's work of the middle 1800's.

Development

Paleontology, and by that we mean the study of fossils, always has been done for a variety of reasons, by persons with divergent philosophical backgrounds. The only dominant, continous intellectual element in the study of fossils has been the theme of deciphering the earth's history. For an inquiry into the meaning of fossils themselves historically we see evolutionary themes, or biogeographic analyses, or an inquiry into the meaning of form as exemplified by fossils, to give a few examples. This helps to account for the diversity of facts and ideas discovered to a significant extent by the study of fossils (*table 1–1*).

Table 1–1. Some major scientific facts or ideas significantly contributed to by the study of fossils.

Physical Sciences

(1) The sequence of events in the earth's history can be determined.
   Example: Climate has changed through time.
   Example: The relative positions of the continents have changed through time.
   Example: Temperature and chemistry of earth's surface have been "constant" for at least the post-Cambrian [post 600 my].

Biological Sciences

(1) The process responsible for new species is organic evolution.
(2) The earth's surface in the geologic past has been divided into regions of different faunas.
(3) Adaptive radiation is a dominant theme in evolution.

By the early part of the 20th century, the general understanding of regional stratigraphy had passed through at least one cycle of interpretation in most of the more highly populated areas of the world. As major discoveries of "field geology—invertebrate paleontology" became fewer, the intellectual frontiers of geological science shifted to other areas.

During this age of exploration, students of invertebrate paleontology were often amateur collectors "whose ambition it was to assemble a cabinet of fine specimens or describe a new species. An amateur would sometimes graduate into a professional; first, by independent publications on a small scale; then, perhaps, by temporary work on a state survey; and finally through recognition of his abilities by appointment to some permanent position on the national survey [U.S. Geological Survey] or in some large museum or university" (Bassler, 1933). The major emphasis was on the description, illustration, and cataloging of the diversity of ancient life.

N. D. Newell put this view in clear perspective in 1954 when he wrote, "It is essential in any assessment of invertebrate paleontology to keep in mind that it originated, and to a large extent has developed, as a stratigraphic tool. In North America, particularly, we have tended to be concerned with the *uses* of invertebrate fossils in the solution of geologic problems rather than with the *meaning*, in the broadest sense, of the fossils."

Newell's comment appears in the book *Status of Invertebrate Paleontology, 1953*. The 14 articles are chiefly summaries of the then prevalent classifications for various groups, and the reasons for those classifications. Although some of the articles continue in the tradition of stating what was described and where it was found, most authors consciously seek to use the descriptive data as grist for the phylogenetic mill. This is a marked step away from the historical association with field geology. A comprehensive phylogeny is a complex hypothesis, and is in the tradition of hypothetico-deductive science already established in Darwin's barnacle monograph (Darwin, 1854; see Ghiselin, 1969a). Many of these 1953 views were later used for the taxonomic organization of volumes in the *Treatise on Invertebrate Paleontology*, published 1953 and subsequently.

By 1968, when the article "Developments, Trends and Outlooks in Paleontology" was published (Moore, *et al.*, 1968), the majority emphasis had shifted to inquiry about how fossil groups "make a living." The question increasingly asked was "How did these fossil organisms use their morphology to conduct the life processes?"

Nevertheless, the fact remains that paleontology is still usually organized around objects rather than ideas. When you ask a paleontologist what he is working on, he will usually say something like "Devonian Brachiopods" rather than, for example, "variation in faunas of different habitats." And hence the history of the study of fossils is different from the development of ideas. The idea of organic evolution, as seen through the eyes of its discoverer Charles Darwin, provides an inspiring example of the development of an idea. Ghislein's (1971) analysis of the successive stages of dealing with this idea is: "In the first place, we have a phase of *illustration*, the goal of which is to become familiar with new ideas, and to establish their utility. Before a scientist even considers embracing a new theory, he needs to understand it, and to see how it can solve problems. . . . A second phase which should tend to come after illustration is quite generally thought to be important: *verification*. One reason why it should come somewhat later is that a test may be very hard to find. . . . After a theory is accepted and tested, there begins a third phase, one which may be called *exploration*. It differs from the illustrative phase, in that familiarity and utility are already granted, and the goal is to find novel applications."

The first stage in the study of objects, such as fossils, is to become familiar with them, particularly to establish their unique properties. This is followed by placing these objects in the context of various generalizing notions. In this second stage of study, fossils achieve importance in proportion to the amount of evidence they provide for or against hypotheses.

At the present time, to judge from our journals, the major effort is still directed toward the description of faunas as a discovery in itself. However, an increasing amount of literature is directed toward what might be called analytical paleontology. This book is directed toward the analytical approach in paleobiology; the method for doing this is to emphasize models.

Models

At this point it is worth mentioning the spectrum of concepts called "models." At one end of the scale is a paradigm which either provides a broad theme that unifies many aspects of a field of scientific research, and in this sense provides a "world-view" of a field, or is a powerful theory or epochal experiment which causes a major body of research to be recast in a new light. As E. O. Wilson has remarked, a paradigm is primarily a psychological event, and its importance is determined by its psychological impact.

As an example of a paradigm in geology, we can cite the Wernerian world-view of neptunism which was used to explain a wide range of geologic features; this was subsequently replaced by the Hutton-Lyell paradigm of uniformitarianism to explain these same phenomena. Kuhn (1962, 1970) has emphasized that when one speaks of a revolution in science, one refers to conceptual changes of such a magnitude that the paradigm dominating a field is replaced.

Other paradigms need not be so inclusive. One reads of "paradigmatic model" or "paradigmatic experiment" in the sense of the optimum and most clear-cut example that makes many other models or experiments understandable. It is in this sense that Rudwick (1961) refers to paradigm to indicate the optimum form of brachiopods of a certain type. Several models in this book are paradigms of this sort.

Generally, the term "model" is used to indicate a concept of lesser scope than a paradigm. But as Hildebrand (1964) makes clear, there are many types of models—from the exact physical model of a planned bridge to a theoretical conceptual model like the ideal gas law. What these and all other models (*table 1–2*) have in common is the deliberate alteration of certain aspects of the real world in order to understand better certain other aspects. The double helix, laboriously first built of wood and metal, and now reproduced by high-school students in cardboard, is useful because spatial relations are maintained. An unimportant variable is discarded in order to express better an important one. The purpose of all models is to provide an abstraction of the real problem into a form that is more manageable for further analysis.

Table 1–2. Some fields of science and discussions of their models.

Field	Reference
Physical and biological science	Kac, 1969
Physical science	Hildebrand, 1964
Geology	Douglas, 1970; Moore, 1967
Population genetics	Lewontin, 1963
Population biology	Levins, 1966
Biology	Stahl, 1967
Biometrics and systematics	Reyment, 1969
Reproductive biology	Ghiselin, 1969c

To derive a model involves a precise methodology: simplification, analysis, and resynthesis. The simple collection of facts in the expectation that an empirical relationship may result does not by itself count as a model since simple induction does not permit manipulation of data. Only by deductively making predictions from a set of data, and testing the consequences of the predictions, does one use a model. A major advantage of establishing a model is to test an idea for major internal inconsistencies before turning to large amounts of data.

In a stimulating paper, Levins (1966) discusses the properties of models. As is obvious from the fact that some aspects of a problem are placed aside in developing another aspect, there is a "cost" involved in model building. Levins identifies the cost as being in either degree of generality, degree of realism, or degree of precision, and notes that only two of these three properties can generally be satisfactorily considered. Invertebrate paleontologists have often been concerned with precision (the particular species), and with reality (the particular conditions under which the species occur), and have been therefore limited in the generality of their observations. Many of the chapters in this book are concerned with generality and applicability in the real world, but lack precision. Models of this type often indicate what data would be useful to collect. Other models, often mathematical, have generality and precision, but lack reality.

Another way to view models is whether they are meant to represent equilibrium (steady-state) conditions, or historical (non-steady-state) conditions in which the system is being perturbed. For example, Worthington (1968) contrasts steady-state models of oceanic circulation with models in which historical climatic fluctuations play the more dominant role. A more decisive example of an equilibrium model is a gas law in which the state of any particular molecule is immaterial to the general description of the behavior of the volume as a whole. The historical view has been the traditional way that biologists and paleontologists have treated their material. However, equilibrium models—where the particular history of species may be immaterial—are gaining interest. The notion that, in some important senses, every species is "equally good" adds generality to a model and often implies the assumption of equilibrium conditions.

In invertebrate paleontology, models that are closely tied to empirical patterns are the rule. The most prominent example of this type of historical model is the view that a plot of number of genera, families, etc., through time, is a true representation of past events. This particular model has now been challenged (Raup, 1972) with an equilibrium model of taxonomic diversity. According to this view, much of the variation in taxonomic diversity is the result of selective bias of the fossil record. What results is a "steady-state" pattern of diversity, rather than continually increasing diversity. Perhaps invertebrate paleontologists would do well to consider other empirical patterns in the light of equilibrium models.

Before one can adequately understand models in paleobiology, it is necessary to have a reasonable idea about the nature of the things being studied. Because of this, an emphasis on biological training has become extremely important. Clearly, morphology, evolution, and distribution heavily rely upon facts and ideas chiefly biological in character. In departments where paleontologists were trained in (and later expected to teach) courses on regional geology, field stratigraphy, structural geology, and even petrology, few of the basic biological notions were considered. The difficulty of this "classical" background was nicely expressed in the early 20th century by Henry Fairfield Osborn.

> The preservation of extinct animals and plants in the rocks is one of the fortunate accidents of time, but to mistake this position as indicative of affinity is about as logical as it would be to bracket the Protozoa, which are principally aquatic organisms, under hydrology, or the Insecta, because of their areal life, under meteorology. No, this is emphatically a misconception which is still working harm in some museums and institutions of learning. Paleontology is not geology, it is zoology or botany; it succeeds only so far as it is pursued in the zoological and biological spirit. (Osborn, 1905: 277.)

Biology presents to paleontology a body of information and approaches which is useful for investigating other phenomena. In this respect, the use of biology in paleontology appears very much like the use of thermodynamics in geochemistry, or classical physics in geophysics. To the extent that problems in invertebrate paleontology are not solvable except by biological approaches, the field is not going to advance unless it realigns itself more directly with ecology, population biology, etc.

Unmixing and Optimization

The evolution of parts of organisms can profitably be considered in terms of maximization in natural selection with respect to external constraints on the system. The chief point is the rather simple one that any time we can put specific phenomena into the context of maximization, then we should look for "unmixing" of phases, i.e., specialization. The effect of maximization will, on the average, tend to be realized to the extent that the stimulus is repeated.

Let us look at a graphic representation of maximization following the technique of linear programming analysis which was used in a more limited context in a study of castes in social insects (Wilson, 1968), and in an analysis of polymorphism in colonial ectoprocts (Schopf, *in press*). We can think of specialization as a contingency plan. The initial step in its development is to conceive of need curves (*figure 1–1A*) for the cases in which two jobs exist. The number of specialists needed to perform job 1 is less than that required if another type of specialist also had to provide for job 1; similarly, the number of the second type of specialist necessary for providing for job 2 is less than would be required if the first specialist also had to provide for job 2. Graphically it can be easily demonstrated that point O_1 in *figure 1–1A* yields the minimum summed number of individuals of the two specialists consistent with effectively conducting both jobs (Wilson, 1968). This point of intersection represents the optimal mix of specialists for the greatest efficiency of operation. This consists of the sum $\alpha + \beta$. We can consider that the production of offspring viable in the next generation is what is being optimized. This is highest per individual specialist at point O_1.

Conclusion: A lower total number of individuals is required if the jobs to be performed can be partitioned into discrete parts. This specialization is the way of all life.

Let us vary the environment. We can conceive of a change in the selection for a different optimal mix of specialists (*figure 1–1B*). This considers the case where additional selection results in the addition of members of specialist 2. The new optimum mix (point O_2) has a larger proportion of specialist 2 than previously was the case. Compare the sum of $\alpha + \beta$ with that in *figure 1–1A*.

Let us continue to change the environment in the same way (*figure 1–1C*). There is so much increased selection for specialist 2 that there is now no point at which a specialist suitable for job 1 would be of selective advantage. The optimum mix (O_3) thus consists entirely of specialist 2.

Figure 1–1: (A) Two jobs being performed by two specialists. The intersection of the "Job 1" line on the N_1 axis is lower than the "Job 2" line so that a lower number of specialists is required to perform Job 1 than

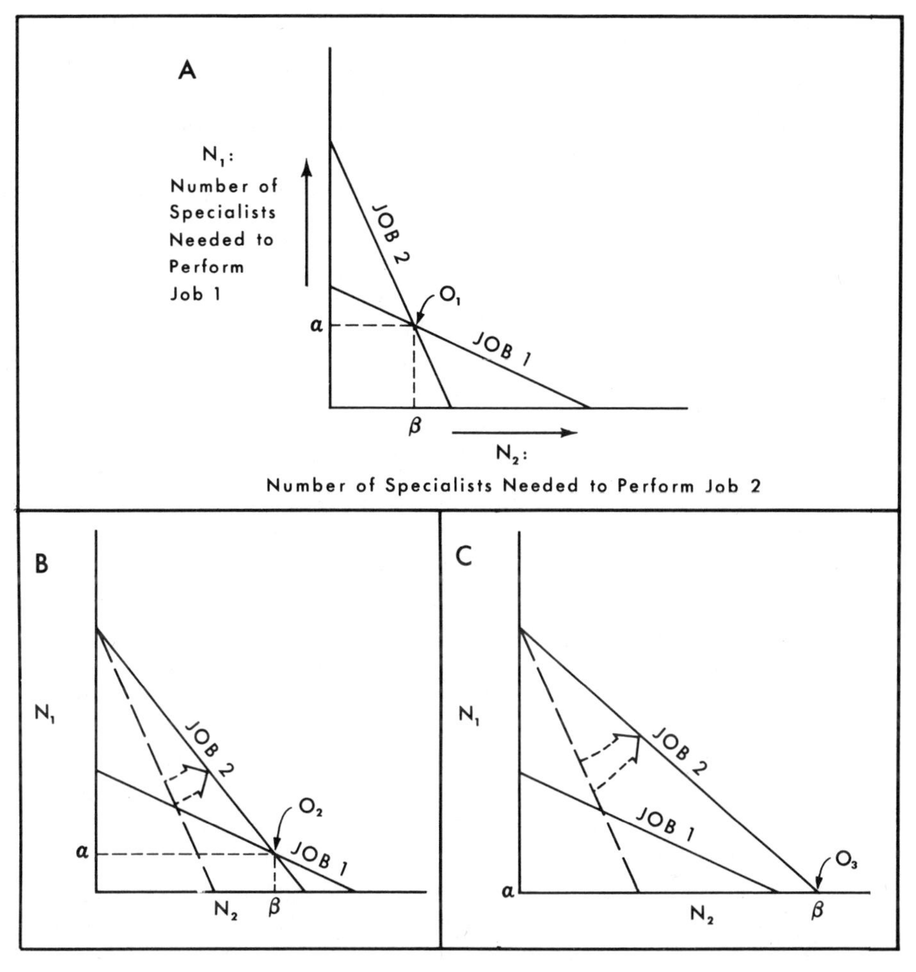

would be required to also perform Job 2. Similarly, a lower number of the second specialist is required to perform Job 2 than would be required if it also had to perform Job 1. The sum $\alpha + \beta$ is the lowest number of individuals of specialists 1 and 2 consistent with performing both jobs 1 and 2. Point O_1 at the intersection of the Job curves is the point of optimization for the smallest number of specialists required to perform both jobs.

(B) Job 2 has become more important resulting in a shift in the equilibrium number of individuals toward more of specialist 2 and fewer of specialist 1. Compare the sum $\alpha + \beta$ with *figure 1A*.

(C) Job 2 has increased in importance to the point that every point on the Job 2 line includes the Job 1 line with the result that there is no advantage to only being able to do Job 1. Specialist 1 disappears ($\alpha = 0$), and β is the number of specialists remaining.

With this simple background, we can summarize the following predictions:

(1) As long as the need for different jobs occurs with relatively constant frequencies, it is an advantage for a species to evolve so that there is one type of specialist to respond to each kind of job. For example, in a non-colonial organism, specialization in shell morphology is predicted to evolve if this results in a partitioning of jobs performed by a non-specialist shell. Similarly, within a colonial organism, polymorphy or castes should evolve.

(2) The countering force which leads to variations in the ratio of specialists to generalists, or even to the disappearance of specialists altogether, is an environment in which the selection pressures are continually changing (*figures 1–1B, 1–1C*). Stability of the selection pressure enhances the evolution of specialists up to the point where the selection pressures are relatively variable for the particular job being performed.

(3) Combining these effects in individuals or taxa with few types of specialization, the ratio of specialists to generalists will be variable. Since a specialist is a more effective performer over a limited range than is a generalist, the specialist will be less abundantly represented.

(4) Variations in degree of specialization can be attributed to the degree of environmental reinforcing of the selection pressure; this morphologic variation can be monitored along stability gradients, specifically latitude, water depth, or evolutionary time. Over time, as a form becomes increasingly specialized in a relatively stable environment, the ancestor-descendent morphological changes should, as a first approximation, roughly parallel latitudinal morphological changes in that same feature. Thus by looking at latitudinal gradients in features we may gain some insight into expected time-dependent relationships.

I noted earlier that this model might be applicable at different levels of biological organization *if maximization can be reasonably expected*, and this is the critical feature.

With respect to a single organism, that which is maximized is the production of its offspring which are reproductively viable in the next generation.

With respect to a single species, that which appears to be maximized is the ability to utilize environmental resources in a way that always tends toward uniqueness; in other words, away from competition. Natural selection will provide an advantage for those species that are not strongly competing.

This analysis may even be applicable at the level of the trophic web or other biological interactions which maintain the partitioning of environmental resources. If so, what *appears* to be maximized is the efficient utilization of available energy resources. In fact, the prediction would be that natural selection will favor the evolution of species which comprise distinct food-webs (i.e., do not compete) which in turn appear to expand to insure the efficient utilization of energy.

Thus one should expect to find (and one does) an increase in both the types of specialization and the degree of specialization within a type, for colonial and non-colonial individuals, for species, and for communities, in those portions of the earth where an individual, a species, or a community can count on environmental stimuli being repeated. This happens to be accentuated in the deep sea and the tropics, apparently to about the same degree. This explanation of diversification has nothing to do with environmental temperature *per se*, food resources *per se*, or any specific environmental factor by itself. Yet it provides an intuitively satisfying conceptual scheme in which one can predict the occurrence and extent of morphologic, organismal, and community diversity.

This heuristic approach must, however, not be confused with considering a species, or a community, as the unit of selection. The notion that communities evolve is ultimately based on a historical statement that communities change through time. When that historical fact is carried over into equilibrium models of communities, the role of individuals becomes lost, and emphasis shifts to energy distribution in the community. In this way, a community has been confused with a "superorganism." Forgotten is the fact that evolution acts at the level of individuals with the result that it *appears* to have maximized higher units of organization. Likewise, succession is a historical statement of a sequence of events, and is not an innate process to be acted upon through natural selection. Whereas the phrases *youth*, *maturity*, *old age*, *rejuvenation*, *overall hoeostasis*, etc. have been heuristically useful in pointing out possible historical trends in subjects as diverse as insect evolution, community biology, and geomorphology, they are also a mirage if taken at face value.

Dimensions

The second general aspect of paleobiology that I want to consider is the question of time scales and size scales of different paleobiological phenomena. These constraints are independent of the significance of any scientific question related to these phenomena. There are four such phenomena whose dimensions I think are worth emphasizing. These are individual organisms, individual species, individual faunal provinces, and individual communities.

The reason for considering the concept of dimensions is that the degree of resolution provided by paleontologic methods must be commensurate with the phenomena being described. Although this fact is intuitively obvious, I am not aware of any previous attempt to make these dimensions explicit. The calculations which follow relate to shallow, continental shelf seas.

Figure 1–2 shows geographic area in square meters on the vertical scale against time in seconds on the horizontal scale.

The shortest time interval on the figure consists of life spans of individuals. For recent organisms, life spans may range from minutes to somewhat more than a thousand years (more than 7 orders of magnitude), but most of the organisms of interest to paleobiologists have life spans of days to years (only 3 or 4 orders of magnitude). The problem for zoologists and botanists in many situations is how to understand the sequence of events in the life span in a *shorter* period of time than the organism normally requires to complete them. From a paleobiological point of view, the problem is different. Individuals of a species that die at different times are mixed together and the question is how to *expand* the fossil record. The resolution in both situations is to measure time-dependent growth factors, such as yearly growth rings.

The sizes of individuals (measured as the area they occupy) also varies considerably, but most are from cm^2 to m^2. Since size is also related to abundance of a taxon (bacteria are many, whales few), the nature of a sampling program is closely connected to considerations of size. Problems of sexual dimorphism in the larger dinosaurs, for example, may be essentially unanswerable because of the paucity of expected finds.

Next, the persistence of species may vary considerably of course, but figures of 1–5 million years per species are not unreasonable. The length of time of a stratigraphic zone has been estimated to be about 1–2 million years in well characterized situations, and these zones are delineated by the evolution of new species. The area occupied by a species during its existence is dependent upon the size, population density, and type of life history. Large, sparsely distributed animals with ease of locomotion occupy more area than small, densely packed forms which are sessile as adults and whose larvae are short-lived. Perhaps areas of 10^6–10^{13} m^2 are appropriate for most modern species.

Marine faunal provinces seem to result from the combination of local geographic features influencing ocean currents in such a way that species tend to be isolated in regions of different temperature. The time scale appropriate to the development and persistence of a distinct faunal province appears on the order of 10–100 million years. Faunal provinces appear to occupy 10^{11}–10^{13} m^2. Faunal provinces may be larger than the ranges of any of the contained species.

Lastly let us consider communities. In one sense, communities such as the level-bottom community, the interstitial community, or the reef community transcend faunal provinces even though the particular species composition may change. In this view, communities have existed as long as particular "general environments" have existed (from 0.1 to 1 billion years) over areas as large as most of a continental region. This is the "equilibrium" view presented as community "A" in the diagram. It is consistent with the views expressed for the other phenomena.

If one views communities from a historical point of view, however, then the particular species composition of a given community is emphasized. In

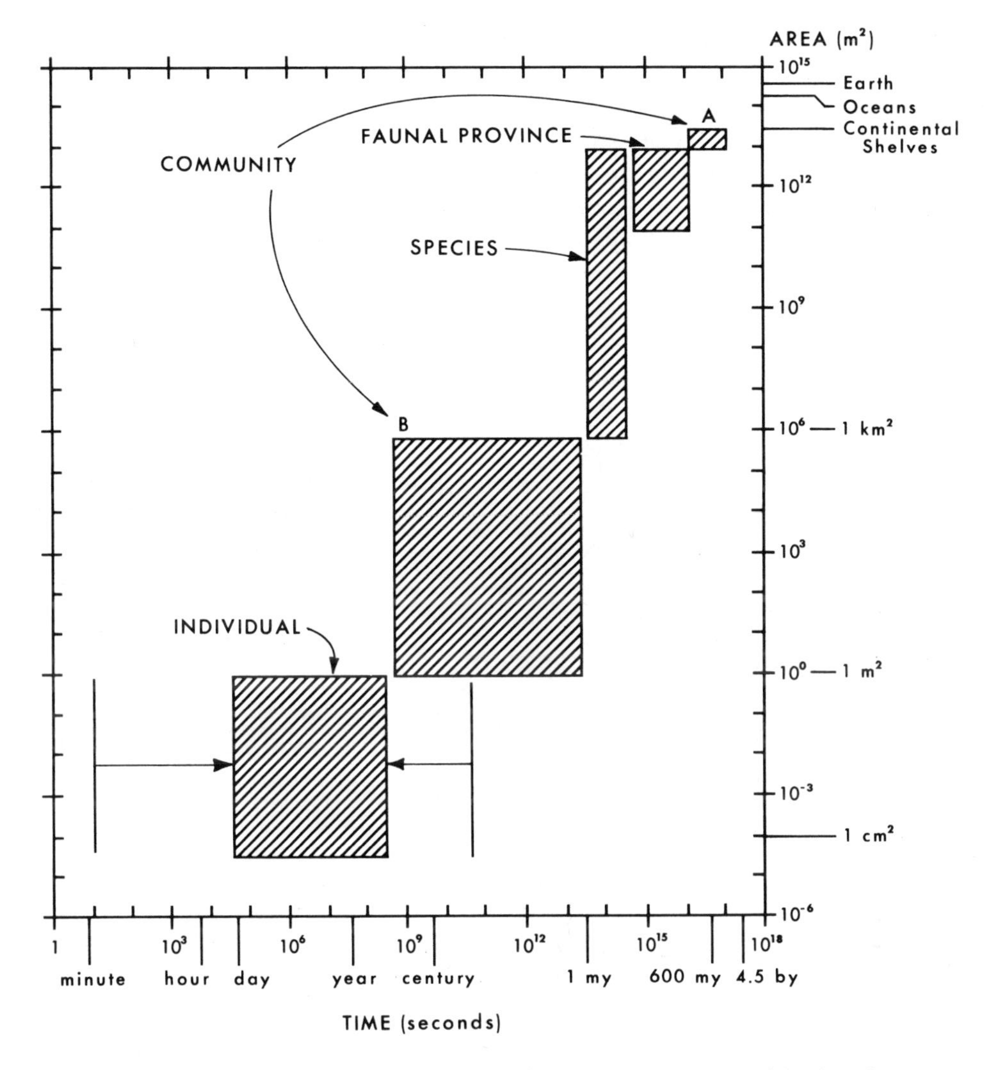

Figure 1–2: Dimensions of paleontologically important items with time in seconds and area in m^2. Community "A" is based on an equilibrium model of communities; Community "B" is based on an historical model of communities. See text.

that case one may refer to several types of level-bottom communities according to the dominant members; these may change from place to place. Such communities would exist on the time scale of the life of that assemblage. This time is greater than that of an individual but less than the time it takes to evolve a new species because, strictly speaking, the substitution of new species changes the community. In area it would be greater than that of an individual but less than that of a faunal province. This "historical" view of communities is presented as community "B" in the diagram.

Since this view of dimensions may be a new point of view, its application and usefulness may not be clear, so let me borrow an analogy from physical oceanography, where this approach is much better understood (*figure 1–3*). Each of the phenomena indicated has its own spectral density in the time-length scale plane. Many examples could be cited where a person set out to

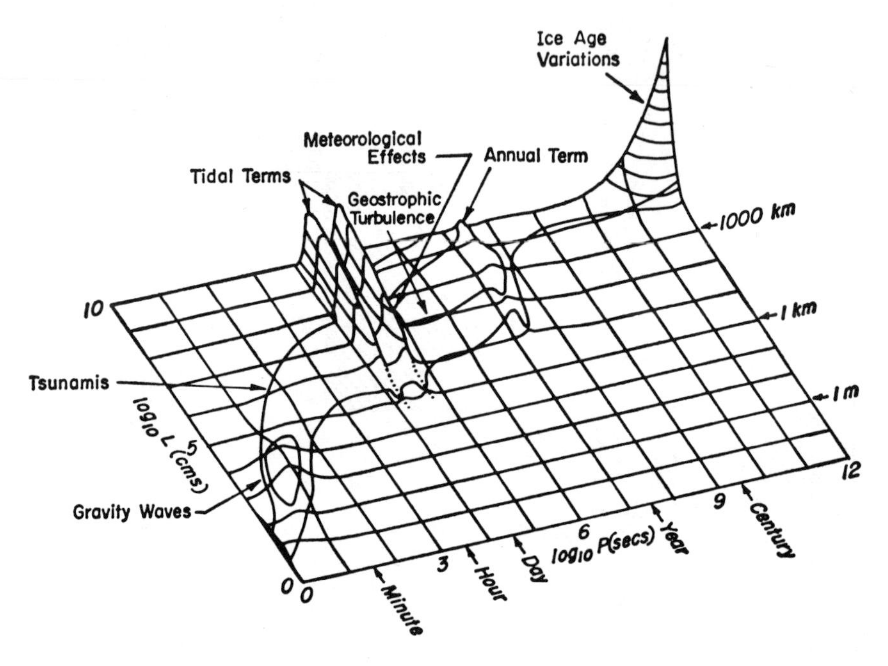

Figure 1–3: Schematic diagram of the spectral distribution of sea level. (From Stommel, 1963, figure 1.) Period, *P*, is in seconds; horizontal scale, *L*, is in cm.

measure phenomena that varied over minutes with instruments designed to measure variation in hours. In paleobiology, as in physical oceanography, one does not explicitly consider as often as is desirable whether a proposed sampling program contains sufficient resolution.

Conclusion

Many will be aware that the subtitle of my paper "Varieties of Paleobiologic Experience" is William James' paraphrased. However, it was Henry Stommel's "Varieties of Oceanographic Experience" (1963) which was, of course, much more the source of my inspiration than James' book. And using paleontology rather than oceanography as the subject matter, I would paraphrase Stommel, and summarize the state of much of our science as follows:

> When paleobiologists set forth to observe and explore the world (even as far as the closest museum), they may be hampered by many adversities, such as faulty locality data, logistics problems, and the absence of fossils of the right sort after arriving at the locality (or museum drawer). However, more often than is commonly admitted, there may be the far more serious handicap of an inadequate design of the expedition as a whole. If we regard an expedition as a scientific experiment, then we must propose to answer certain specific questions, and the strategy of exploration, the disposition of the strata to be sampled, the nature of the collections to be observed, and so on, must be designed with a view to obtaining quantitative, statistically significant answers to these questions. Too often, the design characteristics of paleontologic experiments are such that few biologically or geologically significant answers are obtained.

Appendix. Textbooks of Invertebrate Paleontology

A textbook is directed toward codification of a body of knowledge. Few if any dynamic conceptual advances come first developed in a text. Thus the tracing of the dominant philosophy in these books over a century should indicate the shifts, perhaps even advances, in the "conventional wisdom" of the day. A textbook is also not written until the subject it represents is sufficiently popular to warrant wide distribution. (This was especially so before the days of library exploitation.) Thus the earliest time of a text indicates the period during which the field was sufficiently important to codify the prevailing notions.

In geology, the teachings of Werner (1749–1815) in Germany regarding the Primary, Transitional, and Secondary provided a conceptual scheme for all of geologic history known at that time. Many textbooks were offshoots of this system in the late 1700's and early 1800's. By the mid-1800's, a few books treating chiefly fossils (the definition of a book on paleontology) were appearing. Foremost among these, and one that Zittel (1901) states "was taken as a model for a number of text-books which rapidly made their appearance," was that of François Jules Pictet (1809–1872) which first appeared in 1844–1846.

The second edition of Pictet's *Traité de Paléontologie* of 1853–1857 (first edition not consulted) outlines clearly what paleontology was at that time.

Pictet (1853: 21) begins in volume 1 (of a total of 4) with a brief history of the subject and defines the word "fossil": "*le mot* fossile *sera appliqué à tout corps organisé, enfoui naturellement dans la terre, qui y a été conservé ou qui y a laissé des traces non équivoques de son existence, pourvu que le dépôt dont il fait partie ait été formé sous l'influence de circonstances différentes de celles qui se passent actuellement sous nos yeux*".

After a section on how fossils form, and are altered (diagenesis), he wrote a chapter as long as all of the above (about 40 pages) on the distribution of fossils. He states and then explains 10 laws:

(1) Each species has a limited existence.
(2) Groups of species collectively appear and disappear.
(3) Difference in fossil appearance increases the greater the geologic age.
(4) Diversity in fossil faunas decreases the greater the geologic age.
(5) The more "perfect" the species, the more recent the origin.
(6) The appearance of forms in ontogeny parallels the appearance on earth.
(7) From first to last appearance of a fossil form, there is no interruption.
(8) Temperature has varied over a limited range on the earth's surface.
(9) The geographic range of extinct species was greater than that of living forms.
(10) Physiologic uniformitarianism applies to fossil forms.

Processes responsible for the appearance and disappearance of species are next considered with the qualification (p. 77) that "Mais peut-être aussi la science n'est-elle pas encore assez avancée pour fournir des bases suffisantes à une conviction éclairée."

First Pictet considers extinctions. He believes the most probable of the physical causes to be perturbations which follow dislocations of the crust of the globe. By this he means earthquakes, etc.

The second point is the appearance of species taking the place of those having become extinct. The repopulation of forms from some other land is considered, and rejected for lack of evidence. The theory of the transformation of species is next considered; however, this theory (p. 82) is "complétement inadmissible, et diamétralement opposée à tous les enseignements de la zoologie et de la physiologie." The zoological and physiological principles which call for its rejection stated that if a distinct race did appear as the environment changed, then when the environment changed back, the race would itself disappear.

Having thus been unable to suggest a rational explanation for the appearance of species, he says (p. 87), "Les deux premières explications étant inadmissibles, il reste, la troisième qui est connue sous le nom de *théorie des créations successives*, parce qu'elle admet l'intervention directe du pouvoir créateur au commencement de chaque période géologique." Pictet then immediately admits that special creation is not susceptible to observation, and therefore is not really a theory.

The identification of fossils is next considered with 2 laws stated. First is the law of the unity of organic composition, which says that fossils are (or were) composed of the same stuff as living forms. Second is the law of concordance of parts, which provided the reason why one could assemble a fossil based on modern body plans.

The general sections of Pictet's tome finished with remarks on the use of fossils in stratigraphy and on the classification of geologic deposits. On page 123 he begins with the natural history of particular animal fossils. For each group, the distinguishing characters, distribution, and morphology are cited. And not being constrained by an evolutionary framework he begins with the mammals and in the remaining $3\frac{3}{4}$ volumes works "backward." Volume II begins in fishes and finishes with cephalopods. Volume III goes from snails through clams. And Volume IV begins with the molluscs known as brachiopods and ends with sponges (these being considered lower than foraminifera).

The last 150 pages of volume IV considers the fossils arranged by geologic age, and the implications for geologic history. A fine book of 110 plates accompanied the 4 volumes of text.

An entirely different organization to a paleontology text was followed by Pictet's countryman and contemporary Alcide d'Orbigny. In his *Cours Élémentaire de Paléontologie et de Géologie Stratigraphiques* (1849–1852), the emphasis is on the fossils of each geologic period. In the 4 volumes of text and 1 volume of tables, figures are included with the text as the fauna of each geologic period is developed in sequence. In the initial section, much more attention is given to the process of fossilization than is found in Pictet. The animal kingdom takes about 1 volume to cover, and the stratigraphic relations $2\frac{1}{2}$ volumes.

After the mid-1800's, the major invertebrate paleontology textbooks seem to follow one of 3 lines of organization: stratigraphy (period by period), group by group animal morphology, or evolution.

The period by period approach, possible for d'Orbigny, soon became too unwieldy, although there have been many attempts to revive this approach. A curious descendant is the excellent *The Fossil Record* (Harland, *et al.*, 1967), in which first and last occurrences of major taxonomic divisions of both the plant and animal kingdoms are given.

The group by group approach is best exemplified by the *Handbuch der Paläeontologie* of Karl A. von Zittel, in 5 volumes published over 17 years and not completed until 1893. For each group, the morphology, geologic history, and evolutionary relationships to other groups are shown, with ample illustrations in the text. A translated edition, with considerably rewritten sections by 20 specialists, was issued under the guidance of Charles R. Eastman (1896–1902); a second edition, with 17 major revisors, appeared in 1913. This approach served as the basis for most textbooks in invertebrate paleontology courses in the 20th century.

In England, the standard texts have been by H. H. Swinnerton, Morley Davies, and Henry Woods, all of which are a taxon by taxon approach. Swinnerton's *Outlines of Paleontology* (first edition, 1923; second edition, 1930; third edition, 1949) differs in philosophy and organization from the last 3¾ volumes of Pictet only by arranging the animal kingdom with Protozoa first rather than nearly last. Evolution is accepted and discussed in 10 out of 400 pages. Davies' book *An Introduction to Paleontology* (first edition, 1925; second edition, 1947) deals with groups according to their evident importance in stratigraphy, Chapter 1 being devoted to Brachiopods. The zoological rules of nomenclature are presented, but evolution *per se* is not. Finally, Henry Woods' *Paleontology Invertebrate* (first published, 1893; more than doubled in pages by the 8th edition, 1946) discusses evolution on 3 pages in the Introduction.

During the 20th century in the United States, books on a group by group approach have also predominated. The most famous of these are *An Introduction to the Study of Fossils* (H. W. Shimer, 1914; second edition, 1929), *Invertebrate Paleontology* (W. H. Twenhofel and R. R. Shrock, 1935), later *Principles of Invertebrate Paleontology* (R. R. Shrock and W. H. Twenhofel, 1953), and *Invertebrate Fossils* (R. C. Moore, C. G. Lalicker, and A. G. Fisher, 1952). The approach of all these is identical, as are the texts *Paleontology* (E. W. Berry, 1929), *Invertebrate Paleontology* (W. H. Easton, 1960), and *Course of Evolution* (J. Marvin Weller, 1969).

The goal of all of these standard textbooks is to provide information about the morphology of fossil groups, much as a textbook in organic chemistry discusses the compounds and their morphology.

The third general theme in invertebrate paleontology textbooks is the biological one—the first 100 pages of Pictet. Such a book would be the first 5 chapters of *Eléments de Paléontologie* by Felix Bernard (1895). Fossils are used to exemplify evolutionary principles. In Europe, this general tradition was followed by the well known *Allgemeine Paläeontologie* by Johannes Walther, subtitled *Geologische Probleme in Biologischer Betrachtung*, first published in 1927. In 1959, A. Brouwer published a Dutch *Allgemeine Paleontologie* which was translated into English in 1968. Aspects of distribution and evolution are treated and the book is a true General Paleontology. The excellent recent text by Babin (1971) is also in this tradition.

During the 1950's rumblings occurred on several university campuses in the United States that paleontology textbooks were a set of static classifications and morphologies. One text, *Search for the Past*, by James R. Beerbower (1960), overnight revolutionized for American students the idea of what paleontology is all about. The first 40 per cent is given to a modern and well conceived presentation of how animals lived, and how fossils are useful in solving problems of the geologic record. This was in contrast to about 1 per cent of the space in traditional textbooks at that time. Although many professors didn't adopt Beerbower's text, the students were aware of it, and its spirit caught on. The second edition, published in 1968, began as follows: "The first edition was the somewhat defiant assertion of a young man and of what seemed the culmination of a revolution in paleontology. The revolutionary ardor cools, however, and one begins to wonder if it really was tiger's milk we drank as graduate students. The result: A new edition, polished, perfected, updated, consolidated—and middle aged."

However, at work in other vineyards were those prepared to take the final step, and prepare a book with *no* sections on animal systematics (as about 60 per cent of Beerbower's book remained), and indeed publish a book that deserved the title *Principles of Paleontology*. In 1971, David M. Raup and Steven M. Stanley published this book, whose goal is "to provide a conceptual background for a course in invertebrate paleontology." This emphasis on concepts is exactly what was restricted in other books. This appears to be the book of the 1970's just as Beerbower's was the book of the 1960's.

A point sometimes overlooked, but surely of great importance, is that Raup and Stanley could produce such a book because of the continually appearing installments of the *Treatise on Invertebrate Paleontology* which treat systematics and morphology in extremely well illustrated and thoroughly professionally prepared volumes.

With this type of history, we can look forward to a very rich decade of paleontological research.

PART II

Morphology

2

APPROACHES TO MORPHOLOGIC ANALYSIS

David M. Raup

Editorial introduction. The "science of form," in S. J. Gould's phrase (Gould, 1970a), is a dynamic science emphasizing the functional efficiency of structural designs during the time of development of individuals (ontogeny) and of colonies (astogeny); (superb accounts of various aspects of this field are in Russell, 1916; Thompson, 1942; and Olson and Miller, 1958). Whereas form in this context generally implies external phenotype, there are other aspects of morphology.

In this book, two aspects of morphology are treated—the external phenotype (Raup), and the internal chemical composition (Dodd and Schopf). In these two chapters, the methods of explanation are very similar. The recurring theme is the need to see as contributing to ultimate design *mechanical* (structural or physical-chemical properties), *genetic* (historical-phylogenetic), *ecologic* (ecophenotypic), *functional* (immediately adaptive), and *chance* factors. As models are developed which mechanistically "isolate" one factor while holding others constant, the significance of morphology should become clearer.

Introduction

Approaches to the study of morphology by paleontologists are presently undergoing rapid development. New approaches are being made possible by the appearance of new techniques and analytical methods (especially those involving computers) and old approaches are being reviewed and integrated with modern biological theory.

The present paper has two principal objectives: (1) to summarize and evaluate the current theoretical framework in which morphology is interpreted by paleontologists and (2) to describe the present use and future potential for simulation modeling techniques as applied to morphological problems.

The theoretical discussion makes no pretense of being a complete or exhaustive one. Rather, it is intended to stimulate further thought by pointing to a few problems which are especially critical to the development of morphological analysis—particularly by paleontologists in North America. Fuller discussions of the general subject from a biologic viewpoint may be found in works by Bock and von Wahlert (1965), Mayr (1963), and Waddington (1969), to mention but a few.

At the present time, there is significant and vigorous activity in paleontological research going on in Germany. Much of this work is made possible by a sizeable grant ("Sonderforschungsbereich 53") from the West German government, with the research being coordinated at Tübingen University under the direction of A. Seilacher. This German program is important partly because of its sheer size but more significantly because much of it involves approaches not now emphasized outside continental Europe (see Gould, 1971c). The first half of the present paper is an effort to explore the similarities and differences between the several approaches and points of view.

Much of my discussion will draw on a recent paper by Seilacher (1970) which describes the thrust of one segment (Konstruktions-Morphologie) of the current German research. Also, I am indebted to S. J. Gould for having written his recent review article entitled "Evolutionary Paleontology and the Science of Form" (1970a) because it provides an essential basis for many of the subjects I will discuss.

Points of View

A statement such as "*all morphology is adaptive*" produces little controversy, particularly in Great Britain and the United States. But behind the general acceptance of this statement, and behind the general agreement on the mechanisms of adaptation, lies considerable variety of viewpoint.

Following a series of brilliant morphological analyses written by Rudwick (1961, 1964a, etc.) a significant fraction of invertebrate paleontological research has been directed at interpreting specific morphologic structures in terms of the function or functions they performed for the organisms. Implicit in these investigations has been the conviction that each structure has evolved toward an optimum configuration with reference to its immediate function. In other words, the observed structure is expected to be a close approximation of "the structure that would be capable of fulfilling the function with the maximal efficiency attainable under the limitations imposed by the nature of the materials" (Rudwick, 1964b, p. 36). If the actual structure is *not* a good approximation of the theoretical, optimal structure, it is often assumed that the investigator has incorrectly postulated the function. This general viewpoint is thus what might be called a "hyperselectionist" attitude toward the interpretation of morphology.

One can accept the proposition that all morphology is adaptive and still largely reject the Rudwick approach. It has been argued, for example, that although morphology must be adaptive to the extent that it must function well enough for the organism to live in its environment, the structures of organisms are dominated by aspects inherited from ancestral forms. In other words, a structure such as the cephalopod suture line or the plate pattern in a crinoid calyx may have been modified somewhat by natural selection in order to adapt to specific functional problems, but the total morphology of the structure depends more on phylogenetic heritage than on optimization at the species level. The structures may have originated in the ancestral forms as a series of adaptations (perhaps related to different functional problems than now face the bearer of the structures) or they may represent evolutionary developments relatively independent of adaptation.

The difference in viewpoint just presented is, of course, very close to that discussed by Greene (1958), Rudwick (1964b), and many others, which has been symbolized by Rudwick as the conflict between Schindewolf's "typostrophic" and Simpson's "synthetic" theories of evolution.

Seilacher (1970) has contributed to thinking on this subject by suggesting a framework for interpreting morphology which does not presuppose strong adherence to one viewpoint or the other.

Seilacher's formulation. Seilacher's basic thesis is that any morphologic structure must be looked upon as the result of a combination of forces or influences. He has identified what he considers to be the three most important factors and has expressed them as three corners of a triangle (*figure 2–1*). He further asserts that the principal task in interpreting a given structure is that of placing it within the triangle—and thus assessing the relative strengths of the three factors in that case. The three factors are explained below with examples mostly different from those used by Seilacher.

Historical-phylogenetic factor. This refers to what might be called the "phylogenetic legacy" carried by an evolving lineage. In genetic terms, it would be those aspects of morphology (or morphogenesis) that are not subject to significant genetic variability—either because mutation rates are low or because the structure is controlled by such a large gene complex that rapid modification is unlikely *or* because the gene complex is not subject to allelism.

The historical-phylogenetic factor has the effect of limiting the adaptations possible during evolution. It is thus related to the "Bauplan" concept (see

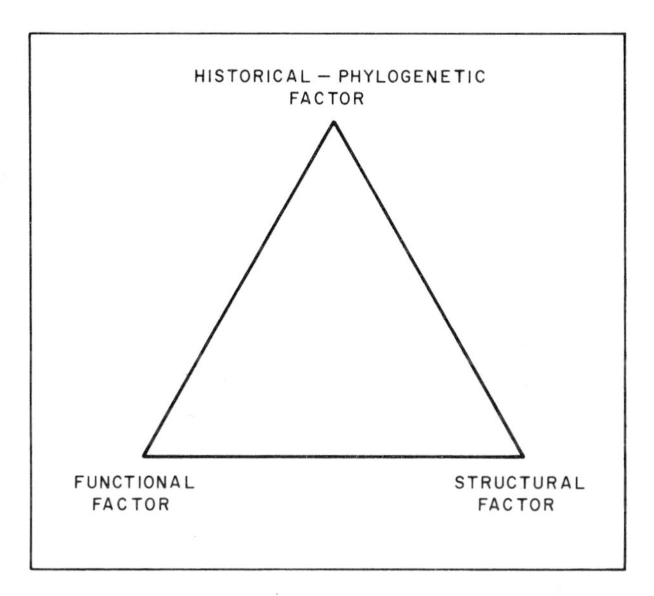

Figure 2–1: Three factors important in controlling morphology. The position of a given morphological feature in the triangle depends on the relative influence of each of the factors. After Seilacher, 1970.

below) of other authors—although formal use of the term "Bauplan" should probably be avoided because it has so many other connotations in the phylogenetic literature.

Example: Most skeletal growth in molluscs (excluding such features as the cephalopod septum) is by gradual accretion of the existing shell. This mode of growth is so fixed or invariant that an evolving molluscan lineage undergoing adaptation is restricted to those kinds of skeletal structures which are compatible with the accretionary system. The shell of the juvenile must be suitable to form a part of the larger shell of the adult. This is in sharp contrast to skeletal growth by periodic molting (as in arthropods), where the juvenile skeleton need be functional only during the ontogenetic stage in which it is deposited.

In nektonic, coiled cephalopods, many of the functional problems are identical to those faced by the naval architect designing submarine hulls (Chamberlain, 1971). Both submarines and coiled cephalopods must solve problems of buoyancy, resistance to hydrostatic pressure, stability, maneuverability, and hydrodynamic efficiency (principally streamlining). But the resulting structures are not even superficially similar because the cephalopod skeleton must grow by accretion whereas the submarine hull has no growth requirements at all.

Functional factor. This refers to the influence of "functional morphology" as this term is generally understood and thus might be called the Rudwick corner of the triangle. It corresponds to those aspects of morphology which result directly and immediately from the process of adaptation through natural selection (not legacies of ancestral adaptations). The paleontological literature provides many examples (by Rudwick among others) of structures which have been successfully interpreted in terms of adaptive solutions to specific functional problems. Gould's review paper (1970a) discusses several of these cases in detail.

Seilacher, by segregating functional and phylogenetic controls at different corners of the triangle, has emphasized that the precise form of an adaptive structure is a resultant of functional factors and phylogenetic restrictions.

Example: Consider again the cephalopod-submarine comparison. Skeletal form in swimming cephalopods is certainly adaptive and may even represent *an* optimum morphology for cephalopods in the nektonic mode of life. Chamberlain (1971) has shown experimentally that the morphology of planispirally coiled ammonoids is consistent with the assertion that evolution in this case has optimized hydrodynamic efficiency.

But, as pointed out above, adaptation in cephalopods took place under certain constraints imposed by inherited morphogenetic systems. The constraint imposed by accretionary growth is but one of many. Thus, cephalopod form does not represent *the* optimum morphology for a swimming organism.

The submarine design produced by the naval architect may or may not be a better solution. The submarine designer also works under historical constraints—meaning a variety of "inherited" specifications for internal instrumentation, propulsion systems, torpedo design, dry dock configuration, simple tradition, and so on.

Structural factor. As the third corner of his triangle, Seilacher is concerned with "non-adaptational elements of low taxonomic significance" (Seilacher, 1970, p. 393). This refers to a variety of morphologic features which are strongly influenced either by physical inevitability, by inherent limitations of materials, or by growth systems which, though necessary to the organism (and therefore adaptive), produce some structures as byproducts which are not necessary.

This category is difficult to define partly because it is heterogeneous and partly because of language difficulties. Seilacher used the term "bautechnischer" and he translated it into English as "architectural" (Seilacher, 1970,

p. 394). I have chosen to use "structural" instead because I have found that many Americans misconstrue the word "architectural" in the present context. A more accurate translation would probably be "constructional" because "bautechnischer" in German usually refers to building construction. A Bautechniker is a building engineer. The term "constructional" must be avoided, however, because Seilacher has referred to the whole triangle approach as "Konstruktions-Morphologie."

Example: Many gastropod shells show spiral elements which originate from irregularities in the shape of the aperture and have the spiral form because of the spiral nature of accretionary growth. Radial structures common among bivalves and brachiopods also reflect the shape of the growing edge. In some cases, the adaptive "purpose" may be found in aperture shape and the spiral structure is inevitable but of no particular adaptive value. In other cases, the reverse may be true: the organism benefits from the spiral structure and produces the irregularities in the growing edge of the shell only because this is the most straightforward way of producing the desired spiral. If both aspects of the structure are adaptive, one is almost certainly more important than the other.

Example: Many skeletons are dominated (in appearance, at least) by mosaic patterns of polygons. Examples include the crudely hexagonal patterns formed by corallite boundaries in colonial corals, plate patterns in echinoderms having rigid skeletons, and the fine structure of many diatom tests. (Gould, 1970a, p. 87, gives additional examples of this phenomenon.) When ancillary features such as echinoderm tubercles and coral septae are ignored, the polygonal patterns are strikingly similar yet the organisms involved are about as distantly related phylogenetically as possible. The similarities thus cannot be attributed to common ancestry (historical-phylogenetic factor). Nor can the similarities be related to common function.

The mosaic patterns are often simply the result of close packing. In the echinoderm and coral examples, the mosaic stems from the fact that the units are of roughly equal size, are equally spaced, and grow in a confined area. Each of the units would have been circular if not constrained by its neighbors. The same is true for the hexagonal pattern of a bee's honeycomb. The morphological effects are similar only because the same physical phenomenon is involved. It is in this sense that Seilacher characterizes structures of this general type as being of "low taxonomic significance."

This is not to imply that the mosaic pattern is independent of genetic control. Clearly, the conditions which make the pattern possible (and even inevitable) are genetically determined. But the details such as the size of angles between sides of polygons, the number of sides, and the lengths of sides are not to be found as coded instructions in DNA.

The mosaic structures are functional in the sense that they are compatible with the functioning of the organism. But they differ from one another in the degree to which their mosaic form is the direct result of adaptation—and thus they differ in their positions in Seilacher's diagram (relative especially to the "functional" and "structural" corners).

In the coral, the mosaic pattern serves to minimize the ratio between corallite boundary area and volume and to minimize the space required for the whole colony. In the diatom example, the pattern forms a natural "sandwich construction"—a design long prized by the structural engineer—which combines light weight and strength (see Hertel, 1966, for discussion).

The functions just noted are undoubtedly real and of importance to the organisms involved but they are by no means the only structures that could have evolved to meet the requirements. Their establishment as parts of the morphology probably stems from the fact that they are easy to develop morphogenetically and call for a minimum of genetic instructions. It is in this sense that they differ from more explicitly adaptive structures such as the zigzag commissure in brachiopods (Rudwick, 1964a). The latter type of structure would be placed closer to the functional corner of Seilacher's triangle.

Example: A prime case of "structural" influence cited by Seilacher (1972) involves various kinds of divaricate sculpture and/or color patterns found in many bivalves and gastropods, in cephalopods, and in some brachiopods. The divaricate structures are characterized by lines which are neither radial nor concentric with respect to the spiral geometry of the shell. Seilacher considers these structures to be essentially inevitable by-products of skeletal morphogenesis and not necessarily adaptive—no matter how striking they may be to the morphologist.

In some instances, a strong case can be made for divaricate features being functionally neutral (Seilacher, 1972)—keeping in mind the difficulties inherent in demonstrating the lack of function for a structure (Rudwick, 1964b). In other cases, the divaricate structures appear to be functional but only as a secondary consideration. In other words, the divaricate structure was used by the organism (perhaps with modification) to solve a functional problem although the adaptation would have been quite different had not the divaricate structure been available.

Yet another kind of "structural" control is related to the inherent limitations imposed on an evolving organism by the nature of the materials and biochemical processes available.

Example: For an organism secreting a skeleton, there is a finite number of materials or compounds that can be used. To cite an extreme case, diamond might be an ideal material for fabricating grinding surfaces (for the mollusc's radula, some vertebrate teeth, etc.). But it is biochemically impossible, as far as is known, for a living organism to secrete diamond. Therefore, secretion of materials less appropriate than the optimum has evolved for the purpose.

Some molluscs have come reasonably close to the diamond by evolving denticle cappings of magnetite—hardness of 6 (Lowenstam, 1962).

Comparison with Rudwick's paradigm approach. Many of the foregoing considerations are included in Rudwick's prescription (1961, 1964b) for analysis of adaptive morphology. The main difference between the formulations of Seilacher and Rudwick is one of emphasis. As noted earlier, Rudwick's approach is dominated by the contention that a given morphological feature should be expected to be very close to the ideal or optimum structure that would be predicted on the basis of its function and within the "limitations imposed by the nature of the materials" (Rudwick, 1964b, p. 36).

It is clear in the context of Rudwick's writings that he means "nature of the materials" to include most of what Seilacher places under the headings of "historical-phylogenetic" and "structural" factors. (Rudwick noted for example that the inherited anatomical relation between the forelimb skeleton and the wing membrane limited the adaptive possibilities for flying in pterodactyls.) But these factors in Rudwick's formulation have been relegated to such a minor status that many followers of Rudwick have tended to neglect them (particularly the phylogenetic factor).

A note on D'Arcy Thompson's approach. D'Arcy Thompson's *Growth and Form* (1942) has rightfully had an enormous and beneficial influence on paleontological approaches to morphology. The irony in the present context is that Thompson emphasized the functional and structural aspects of morphology to the exclusion of everything else. To quote from Medawar (1967, p. 25), "... D'Arcy was an anti-Darwinian! Believing as he did that present phenomena [observed morphology] should be explained by present causes, he saw the appeal to deep historical antecedents [phylogenetic legacy] as an evasion of responsibility ..." In terms of modern evolutionary biology, Thompson was contending that the genetic makeup of an evolving organism is entirely thus plastic, that it can be altered completely as part of species-level adaptation to immediate functional problems.

Additional Factors Influencing Morphology

The implication of the foregoing discussion is that if the three principal factors (*figure 2–1*) are understood and if their relative importances in a given case are known, then the morphology which evolves can be predicted. This is probably not true even under ideal circumstances and therefore other factors must be playing an important role in the determination of morphology. The two most important seem to be *chance* and *ecophenotypic* effects.

Chance in adaptation. Attention to chance factors becomes important in several ways in the analysis of specific morphologies. Consider, for example, an evolving lineage adapting to a specific functional problem but a problem which has several solutions even under the existing phylogenetic and structural constraints and opportunities. One of these solutions may evolve rather than the others—not as a result of a selection process but by simple chance.

This can be illustrated by the diagram of adaptive peaks and valleys (à la Sewell Wright, 1932) shown in *figure 2–2*. The coordinates are genetically

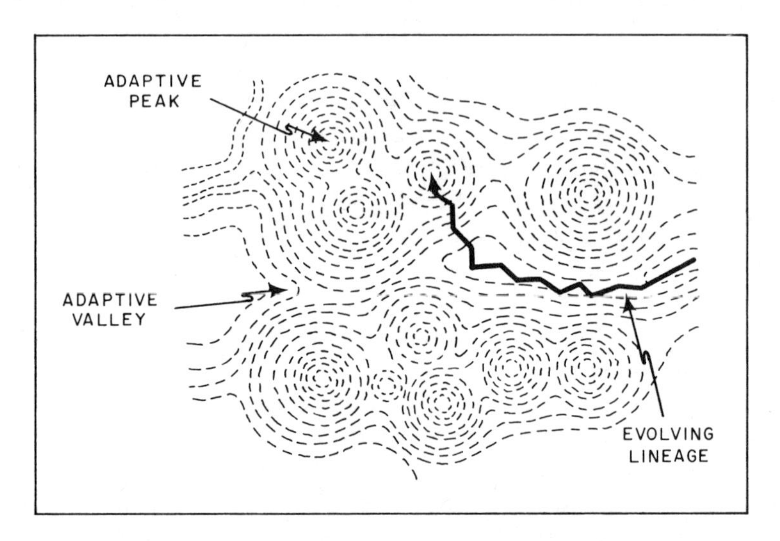

Figure 2–2: Adaptive surface showing the course followed by the evolution of a hypothetical lineage.

controlled morphologic parameters so that the contoured area represents all morphologically possible structures (keeping in mind that a real case would generally be much more complex and require many more dimensions). The topographic highs are adaptive peaks, or optimal solutions.

The evolutionary course followed by a hypothetical lineage is also shown in *figure 2–2*. It climbs one of the smaller adaptive peaks. It does not go down the other side because to do so would be to evolve toward a less well adapted form. The only ways the lineage could get off the first peak are (a) to have the peak "move out from under" the lineage—perhaps through geologic change of environmental conditions—or (b) if populations occupying the peak have genetic variability great enough to include higher slopes of adjacent adaptive peaks.

That both of these "escape" mechanisms are used is obvious from the fact that evolution did not long ago cease. But it must also be true that many species become fixed on some adaptive peaks while other perhaps higher peaks remain vacant and that the difference is a chance process. This means in turn that the morphology we observe in a given case is often not the best possible under the "limitations imposed by the nature of the materials."

Several authors have searched for major vacant adaptive types. Hertel (1966), for example, has noted that wheeled vehicles have not developed in the biological world. He presents several convincing arguments for the difficulty of adapting the concept of the wheel for locomotion in a living organism. But most of his arguments are based on the assumption that the wheel is to be used to transmit locomotive power as well as to reduce friction. If we require only the latter function (as in the wheels of an airplane or any vehicle which has an independent means of propulsion), then the lack of wheeled organisms is not so easy to justify.

Another example may be found in the lack of organisms using sails and wind power to travel over the surface of water. Though this does occur in a very primitive form in certain jellyfish, it is surprising that a fully sophisticated system (with full maneuverability) has not evolved—unless the role of chance has been significant.

Thinking on this subject is hampered by the inevitable tendency to accept what *has* evolved as natural and to reject what *has not* evolved as impossible or impractical. It is doubtful, for example, that flying insects or jet propelled cephalopods would be thought likely were it not for the fact that they did evolve.

Ecophenotypic effects. It is common knowledge that the phenotypic expression of a given genotype is often strongly influenced by the environment, the range of morphology (phenotypes) possible under one genotype being referred to as the "norm of reaction." Because ecophenotypic effects often play a large role in the morphology one observes, this factor cannot be ignored in morphological analysis.

For example, consider the effects of population density on form in such organisms as oysters. The differences in morphology between the crowded and uncrowded may be substantial and overwhelmingly significant statistically.

The position of the apical system in the sand dollar *Dendraster excentricus* appears to be strongly dependent on exposure to high energy environments (Raup, 1956). Individuals develop a highly eccentric apical system or not, depending on environment, but without genetic differences.

Another kind of example is the well known effect of environment on isotope ratios (O^{18}/O^{16}) and trace element contents (Mg and Sr in biogenic carbonates, for example). Genetically identical individuals show different skeletal compositions depending on environment.

It should be noted that in almost all cases of ecophenotypic variation, the same kinds of differences are known to occur as genetically controlled phenomena in other species (as, for example, taxonomically correlated isotopic variation) (Weber and Raup, 1966b).

Morphologic features produced by or dominated by ecophenotypic effects are not interpretable in terms of any of the three corners of the triangle in *figure 2–1*—except to the extent that the degree of phenotypic plasticity (the size of the norm of reaction) may be considered advantageous to the species. Some ecophenotypic effects are clearly functional while others are not. The sand dollar case is functional in that the extreme position of the apical system is of benefit to the organism in its feeding position under conditions of turbulence. Differences in isotope ratios and trace element contents, on the other hand, have no known functional significance. The morphologic effects of crowding are probably detrimental to the individual: oysters, for example, undoubtedly function better and shell form is more nearly optimal when they are living in an uncrowded condition.

Many more examples of ecophenotypic effects could be given (see Hubbard, 1970, for a recent paleontological example concerned with morphology of corals). The literature contains hundreds of examples—ranging from effects of nutritional deficiencies that dominate an organism throughout its development to effects which result from occasional accidents in ontogeny that modify later development. Sometimes whole populations are affected uniformly; sometimes the effect is limited to a few individuals. In either case, these effects form an important aspect of the morphology seen by the paleontologist. A particularly striking example is the occasional departure from planispiral coiling in ammonoids in order to compensate for imbalance caused by an attached oyster (Merkt, 1966).

Aids to the Analysis of Morphology

Figure 2–3 shows the array of factors controlling morphology that have been discussed in the foregoing sections. Any study of morphology must inevitably grapple with the problem of assessing the relative importance of the

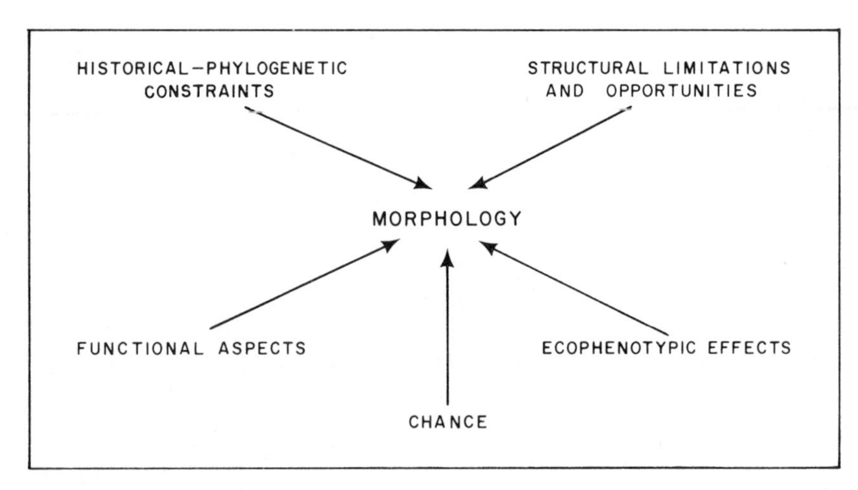

Figure 2–3: Principal factors controlling morphology.

several factors. Such apparently straightforward problems as defining taxonomic characters and recognizing homologies depend on correct recognition of causative factors. Purely ecophenotypic variation must be recognized, for example, before meaningful taxonomic characters can be assigned. If similarities due to structural phenomena, such as the mosaics discussed earlier, are not recognized false homologies may be established.

Several logical aids or analytical techniques have been developed in recent years to facilitate morphological analysis. Factor analysis coupled with clustering techniques have, for example, made it possible to reduce masses of morphologic data to a basic set of variables. Most of these developments are amply discussed elsewhere and will not be dwelled on here. There is one, however—simulation modeling—which is relatively new and which deserves expanded discussion, particularly because it may in future be more valuable than some of the others in unravelling the complexities implied by *figure 2–3*.

Simulation Models

The term "model" is widely used and abused. I will confine myself in this paper to models which simulate morphology or simulate an aspect of the functioning processes of an organism in its environment. I will use the classification of simulation models contained in Harbaugh and Bonham-Carter (1970), as shown in *table 2–1*.

Physical models are actual laboratory models which are most often used to study some aspect of the mechanical operation of the morphology being modeled. They may be constructed at actual scale or increased or decreased in size as convenient to the problem. Study of the behavior of replicas of ship hulls in flumes or tow tanks is a common use of physical models. Schmidt (1930), Kummel and Lloyd (1955), and Chamberlain (1971) have all used specimens or models of coiled cephalopods in this way to study the shell's hydrodynamics.

Symbolic models are usually mathematical but other symbolism may be used. In either case, the morphology or some function of the morphology is simulated in a symbolic form. Simulation of ideal spirally coiled shells by computer (Raup and Michelson, 1965) is an example of a mathematical symbolic model of morphology.

Dynamic models are ones where there is some "feedback" between output and input where a process is being simulated. Dynamic models are commonly used by sedimentologists to simulate basin filling. They are dynamic in the sense that simulated sedimentation at any stage of basin filling affects the character of subsequent sedimentation. In paleontology, simulation of meander tracks of grazing organisms (Raup and Seilacher, 1969) is based on a dynamic model to the extent that the simulated organism is programmed to avoid its earlier track and thus its movement at a late stage in track development is influenced by earlier stages.

Static models lack feedback. They may simulate a time-dependent process —as, for example, the symbolic modeling of coiled shells simulates accretionary growth with a greatly compressed time scale—but the outcome is not influenced by intermediate output.

Deterministic models contain no random elements. That is, the outcome of the simulation is fully determined by the input parameters. This means that successive "runs" under the same initial conditions using a deterministic model yield identical results. This will be true whether the model is static or dynamic.

Probabilistic models (stochastic models) contain at least some random elements. The grazing track example is based on a probabilistic model in that the lengths of some meanders are selected by a chance process. Inevitably, repeated "runs" using the same conditions yield different results. The amount of variation in such results depends on the amount of randomness built into the simulation.

As can be seen in *table 2–1*, eight different kinds of models can be produced on the basis of the foregoing distinctions. The grazing track simulation is thus called a dynamic-symbolic-probabilistic model.

Simulation models in morphologic studies. *Table 2–1* is filled in with as many paleontological examples from invertebrates as are possible within the scope of purely morphological studies. If the scope were widened to include all paleontology, there would, of course, be more citations. In taxonomy, construction of dendrograms from numerical taxonomic data would exemplify a static-symbolic-deterministic simulation of phenetic relationships. In taphonomy, experimental burial of molluscan shells in a current (Clifton and Boggs, 1970) would be a physical-static-deterministic simulation, as would the experimental aspects of Reyment's classic work on post-mortem distribution of cephalopod shells (1958).

It should be noted that the distinctions between the several kinds of model are often difficult to make. The shell burial example could be considered dynamic rather than static because the initial stages of the burial process certainly have a profound influence over subsequent stages. The categories in *table 2–1* should be looked upon, therefore, as expressing a *range* of possible simulation models rather than a set of perfectly discrete types.

Table 2–1 makes clear that the vast majority of simulation modeling done so far with morphological problems has involved physical models in a static-deterministic framework. And of these, most have been designed to test the functional efficiency of a particular form or structure. Rudwick (1961) constructed working models of the brachiopod *Prorichthofenia* to test a postulated function for various morphological features related to feeding mechanisms.

Table 2–1. Paleontological uses of simulation models in morphological studies.

	Deterministic Models	
	Physical	Symbolic
DYNAMIC	Growth of echinoid tests using soap bubbles (Raup, 1968)	Plate growth in echinoids (Raup, 1968)
STATIC	Mode of life in *Calceola* (Richter, 1929)	Form of coiled shells (Raup and Michelson, 1965)
	Hydrodynamics of coiled cephalopods (Schmidt, 1930; Kummel and Lloyd, 1955; Chamberlain, 1969, 1971)	Form of planktonic foraminifera (Berger, 1969)
	Hydrostatics of straight cephalopods (Tobien, 1949)	
	Feeding in *Prorichthofenia* (Rudwick, 1961)	
	Mode of life in *Posidonia* (Jeffries and Minton, 1965)	
	Mode of life in spiriferid brachiopods (Wallace and Ager, 1966)	
	Mode of life in *Gryphaea* (Hallam, 1968)	
	Mode of life in *Kochiproductus* (Shiells, 1968)	
	Formation of cephalopod septum (Seilacher, pers. comm.)	
	Probabilistic Models	
	Physical	Symbolic
DYNAMIC		Meander trails of grazing organisms (Raup and Seilacher, 1969)
STATIC		

Kummel and Lloyd (1955) used plaster casts of actual fossil cephalopods to test (in a moving current of water) the streamlining characteristics of different shell forms (see also Wallace and Ager, 1966). Chamberlain (1969, 1971) had the same objectives but used computer-produced scale models of coiled cephalopods and propelled the models through still water.

All of the models mentioned in the preceding paragraph are obviously directed at assessment of the functional corner of Seilacher's triangle (*figure 2–1*) and all use what is basically Rudwick's paradigm approach. In each case a function was postulated, a model of morphology constructed, and its functional efficiency tested empirically.

In contrast, Seilacher's membrane model for septal shape formation in shelled cephalopods (Seilacher, personal communication) is an example of a static-physical model used to assess the structural factor in morphogenesis. He suspended uniformly elastic membranes from wires that had been bent to the shapes of actual suture lines of fossil cephalopods in order to test the theory (Thompson, 1942) that the septal surface is a surface of minimum area attached along the suture line. The test in this case was a failure because the experimentally produced surfaces were *not* good approximations of actual septal surfaces. But Seilacher's study is valuable as an example of effective use of simulation modeling just because of its failure. It served to exclude an explanation of septal shape which prior to testing had wide acceptance. The membrane experiments opened the way for new propositions concerning the morphogenesis of the septal surface which can in turn be tested by modeling techniques.

As can be seen from *table 2–1*, most categories of simulation modeling (except for the static-physical-deterministic) are either devoid of examples using fossil morphology or nearly so. It is interesting to note that, in other fields, the distribution of applications is quite different (see Harbaugh and Bonham-Carter, 1970). In sedimentology, for example, most modeling to date has been of the dynamic type (either physical, using flumes, or symbolic, using digital computers). In the discussion that follows, I will explore some of the untapped areas for modeling in the paleontological study of morphology.

Future applications of simulation modeling. The search for morphologic problems for which simulation modeling offers particular promise of being able to solve problems not otherwise soluble can be based on the expanded set of controlling factors developed in this paper and shown in *figure 2–3*. In what ways would modeling be helpful in isolating and interpreting the several controlling factors? We will look at each of the factors in turn and assess the potential for successful modeling.

Historical-phylogenetic factor. At first glance, this factor does not seem amenable to modeling analysis by virtue of its complexity and lack of inherent order. In special cases, however, the phylogenetic constraints under which an organism evolves can be isolated and the effects simulated.

Simulation of the general form of coiled shells is an example. When the helical, equiangular spiral is used to simulate (symbolically, by computer) a spectrum of "possible" forms one is, in effect, simulating those aspects of shell form which are subject to phylogenetic constraint. That is, one is assuming that the shell must grow as an expanding, hollow tube (by accretion at its open end) and that growth gradients are such that the tube follows an equiangular spiral. These conditions represent the "Bauplan" that the organisms are operating with in adaptive evolution. Individual lineages are "free" to modify the parameters of the spiral (such as reducing the helicoid parameter to zero to make a planispiral shell) and may even in some cases alter the basic spiral (see Kullmann and Scheuch, 1970, for example). The "Bauplan" presumably originated as an adaptation in early molluscs, etc., and it can conceivably be replaced by something totally different through adaptation. The fact that it has been remarkably stable during the evolution of groups of organisms employing it suggests that it represents the summit of an adaptive peak from which it is "difficult to get off."

If the foregoing interpretation of the phylogenetic constraints of organisms with coiled shells is correct, then the computer simulation of all possible forms provides the array of forms which are selected from for adaptation to specific modes of life or physical environments.

One of the greatest values of the simulation exercise is that it tests one's understanding of phylogenetic constraints. If a model is postulated for the shape of the ammonoid suture, for example, but, when implemented by a computer program, *fails* to produce convincing simulations of several important sutural types, one can only conclude that the postulated phylogenetic controls are incomplete or have not been fully understood.

Functional factor. As we have seen, most morphological modeling in paleontology has been directed at determining the functional significance of morphologic features. It is not coincidental that these have been based on physical rather than symbolic models. The principal reason for this is that most of the problems are not amenable to mathematical solutions and thus must depend on empirical data from scale model experiments. The same is true of course for many comparable engineering problems (airplane design, bridge design, etc.).

Structural factor. Modeling mosaic structures with soap bubbles (Raup, 1968), close-packed elastic spheres (Helmcke and Otto, 1962), and the like are examples of study of structural factors through simulation modeling. D'Arcy Thompson's *Growth and Form* contains abundant examples of other possibilities. Many of these, such as his models of septal insertion in corals, have obvious paleontological application and should be investigated further.

The greatest value of such studies lies in the discovery that structures which appear extremely complex are reducible to the result of simple physical or chemical processes requiring many fewer descriptive parameters than the number of taxonomic characters normally applied to them.

Chance factor. With the exception of the grazing track simulation (*table*

2–1), none of the models discussed thus far contains chance elements. Yet we have seen that chance is an important determiner of morphology, at least to the extent that chance plays an important role in the evolution process.

Simulation of evolution using stochastic models has been the subject of some work by population geneticists but these studies have been directed at purely numerical investigations of gene frequencies under conditions of simulated selection and mutation. The same sort of approach could also be applied to the morphological problems being considered here. Given a phylogenetic legacy or "Bauplan," certain structural limitations or possibilities, and given functional objectives, it might be possible to simulate the process whereby one or more of several possible morphologic solutions are "selected" by evolution.

Ecophenotypic effects. Direct environmental modification of morphology can be modeled by many of the techniques already discussed. The possibilities for probabilistic models are particularly great in this context. Consider, for example, the external form of a solitary tetracoral. The skeleton of the adult individual carries a record of many events during ontogeny—most of them representing irregularly spaced incidents of ecophenotypic modification not unlike the variation in thickness and other characteristics seen in a tree-ring sequence. This sort of morphologic feature is extremely difficult to deal with because of the randomness of many of its attributes. A probabilistic simulation model which successfully segregated the genetically controlled (deterministic) aspects from those governed by chance might make such morphology much more readily interpretable. The result would be a much clearer picture of the distinction between taxonomically important and paleoecologically important attributes.

3

APPROACHES TO BIOGEOCHEMISTRY

J. Robert Dodd · Thomas J. M. Schopf

Editorial introduction. There is a strong and prevailing antithesis to many of the "results" of the study of the chemistry of fossils (biogeochemistry). This feeling is summarized by Raup and Stanley (1971): "Our omission of topics that might together be called 'biogeochemistry' may draw criticism, but we believe that many isotopic and trace-element approaches undertaken during the past two decades have thus far contributed little to general paleontologic knowledge, largely because of the thorny problems imposed by diagenetic alteration." And they therefore did not include in their text even their own work in this field.

Dodd and Schopf wish to place the emphasis, not on diagenesis, but on a different area, namely the relative lack of carefully controlled experiments in biogeochemistry. The paleontological way of doing things has been to apply rather than to learn. To get away from this, they draw attention to thinking about biogeochemical problems in terms of partitioning the initial influence on the skeletal composition into physical-chemical, genetic, or environmental factors. Obviously diagenesis is a problem, but this falls into the realm of choosing the right material, and is not part of the experimental design after the material is obtained.

In their examples, Dodd and Schopf emphasize elemental, mineralogic, and oxygen isotope applications in marine invertebrates. The methodology they support should be applicable to the increasing use of organic geochemistry in paleontological problems, so far best seen in paleobotany. Rather than champion any specific model, these authors try to present a way to look at biogeochemical problems which necessarily will utilize the model approach: simplification, analysis, resynthesis.

Introduction

In science there are often two basically different ways to view relationships: either one traces the development of diversity or one attempts to find general, simplifying explanations. This conflict of the particular versus the general appears in many fields of biology (Lewontin, 1969a), three of which (biogeography, community biology, and biogeochemistry) are discussed in this book. These opposing viewpoints are perhaps more clearly defined in biogeochemistry than in any other field of paleontology. Many have generalized from equilibrium chemical processes while others have stressed the variability of the chemistry of skeletal materials.

Although a considerable body of data in biogeochemistry had been obtained prior to the early 1950's (summarized by Vinogradov, 1953), it was not until then that modern research came into its own. The most important single stimulus to the field was probably the development of the oxygen isotopic paleothermometer by H. C. Urey and his co-workers (Urey *et al.*, 1951). Much of the research since then has been physical-chemical and has emphasized explanations in terms of inorganic equilibrium precipitates. But much of the variability of skeletal chemistry has not been accounted for by this approach.

The genius of Darwin and of Mendel, and indeed the source of many achievements in molecular biology, has been in recognizing that the signal was in the *variation*, not in the *average* of the variation (Lewontin, personal comments; Ghiselin, 1971). However, in biogeochemistry, paleontologists have often been seeking principles analogous to the ideal gas laws—that is, generalizations due to average behavior of elements. Thus if Mg increases with temperature in the calcitic skeletons of some organisms, then why not in all such organisms? Perhaps a more useful approach would be to study the variation observed in order to understand the biological "message" it contains.

The explanation of the distribution of morphological phenotypes (e.g., clams with long siphons) is readily perceived in terms of natural selection. No one claims that all clams must have long siphons in order to appreciate the reason why some of them do. No such approach has yet indicated why Mg does not *always* increase in calcite with temperature. In other words, the modification of the physical-chemical system is not yet displayed in terms of theory convenient to paleontologists, most importantly in terms of evolutionary theory.

Most biogeochemical studies to date have been either simple observations and reporting of factual data or attempts to use these data for the solution of problems (especially environmental interpretations). Few studies have been specifically designed to study the fundamental factors actually controlling skeletal chemistry. Few attempts have been made at proposing and testing models to explain observed biogeochemical relationships. Indeed the nature of the problems of biogeochemistry has not always been made clear. Perhaps the precision of the chemist and his tools lulled some into expecting that resolution would result in meaning, following the converse of the dictum that if you cannot measure it, it does not exist.

The General Problem

We believe that the chemistry of an organism can be profitably partitioned into three independent primary causal factors (physical-chemical, genetic (biochemical), and environmental) with a fourth factor (diagenesis) of great practical concern in deciphering original composition (*figure 3–1*). As an example, of three species of ectoproct living on a mussel shell at Woods Hole, one may be aragonite, another calcite, and the third composed of both minerals, thus showing a genetic influence on composition. The amount of Mg in the ectoproct calcite may increase along a temperature gradient, thus

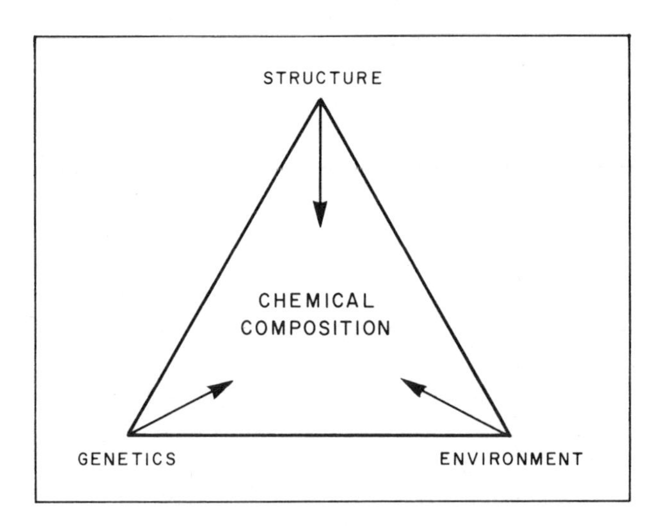

Figure 3–1: Diagrammatic representation of the primary controls of original chemical composition.

showing an environmental influence. Because of crystal structure limitations and the smaller ionic radius of Mg^{++}, that element may be more abundant in calcite than in aragonite, illustrating the influence of a physical-chemical factor on elemental composition. If the specimen were fossilized we might well find that the originally aragonitic portion of the ectoproct had been replaced by calcite, showing the effect of diagenesis.

Three of these components are similar to those used in the analysis of morphology in the chapter by Raup (physical-chemical = structural, genetic = phylogenetic-historical, and environmental = ecophenotypic). The "functional" significance of composition is almost never examined, perhaps because of its relative uniformity. Further, "chance" factors are rarely mentioned. The central problem is to establish the control and range of variation caused by physical-chemical, genetic, environmental, and diagenetic influences on the chemistry of fossils.

How successful have past studies been in revealing the relative and independent importance of these four factors? Much of the effort that has been expended in trying to understand the chemistry of fossils has emphasized physical-chemical influences. To a certain extent, environmental variability is simply a matter of physical-chemical influences over a range of temperatures and water chemistries, and these are then considered the first order influences. But to a large extent, the approach has been an empirical one with the emphasis on using a derived relationship to estimate paleosalinity or paleotemperature rather than on understanding the basic mechanisms involved in controlling skeletal chemistry. Other environmental aspects have been considered second order phenomena. Perhaps the genetic influence has been the least appreciated. Although empirical differences between different species have been long noticed, the explanatory models to link variation in skeletal chemistry and genetic influences have been little studied. Finally, several of the earlier studies did not appreciate the significance of diagenetic effects on altering original composition.

Part of the difficulty in evaluating genetic and some environmental factors has been due to the nature of the samples being studied. All too often biogeochemical studies have been based on a phylum by phylum study of museum specimens or material that just happened to be available. Carefully selected material chosen to represent different biological properties may yield much more valuable results. The genetic and environmental factors should be especially susceptible to the experimental approach.

A different but still fundamental question is "What part of the chemical variation is adaptive?" Does it really make any difference to a species that it secretes calcite or aragonite, or both, or apatite, or quartz? The problem in this case is to partition the opportunities and limitations of genetic, environmental, and physical-chemical factors in order to assign some sort of selective advantage or disadvantage to the particular chemical characteristic measured.

Another possibility is that natural selection may not be acting directly on the chemistry of the skeleton but rather on some biochemical or physiologic process of the organism which is only secondarily related to chemistry. Finally, the extent to which chemical differences are due to pure chance must also be considered. If one understood the reasons for composition framed in terms of natural selection for different chemistries of hard parts, there would be added incentive to learn whether fossil groups (such as tabulate and rugose corals, stromatoporoids, nautiloids, brachiopods, etc., of the Paleozoic) were originally calcite or aragonite.

The three primary factors in our simplifying model are not totally independent, of course. For example, differences in skeletal chemistry between different taxa would be classified as caused by a genetic factor but might also be explained in terms of physical-chemical differences within the organism at the calcification site. Variation of skeletal chemistry with an environmental factor such as temperature might in some cases be caused by genetic differences in organisms between high and low temperature regions. And some environmental differences (for example, the temperature effect on the O^{18}/O^{16} ratio) can be explained in physical-chemical terms. Thus in the following sections some examples that are discussed under physical-chemical factors could be considered environmental and vice versa.

In the following discussion we briefly present some of the more prominent aspects of variation in composition due to physical-chemical, genetic, environmental, and diagenetic influences, and the models that have been proposed to explain them. Under each section (except diagenesis) we give examples of how mineralogy, chemistry, and oxygen isotopic composition are influenced.

Physical-chemical Factors

Aragonite is not a stable mineral under surface temperature-pressure conditions (Fyfe and Bischoff, 1965; MacDonald, 1956). Calcite has a slightly smaller free energy of formation than aragonite (-269.7 kcal/mol vs. -269.9 kcal/mol at 25°C; Robie and Waldbaum, 1968) and aragonite is more soluble (solubility product $= 4.7 \times 10^{-9}$ compared to calcite with solubility product $= 6.9 \times 10^{-9}$, from Latimer, 1952). Thus if the mineralogy of carbonate skeletons were controlled only by physical-chemical factors, they should be calcite. However, calcium carbonate that is inorganically precipitated from sea water is usually aragonite although high-Mg calcite may form under special conditions (Fischer and Garrison, 1967), and aragonite is certainly common in skeletons of organisms.

The question posed is "Why does an unstable mineral (aragonite) form inorganically in sea water as well as in skeletal tissues?" "Equilibrium" theories to explain this are related to crystal poisons or "cation interference," on the one hand, and "organic interference," on the other. "Historical" or natural selection theories have not been proposed.

Laboratory precipitation experiments and field studies by many different workers (e.g., Murray, 1954, and Kitano, 1962) have shown that the concentration of several ions has a strong influence over whether calcite or aragonite is formed (*figure 3–2*). The most effective ion in controlling mineralogy seems to be Mg^{++}. At low Mg^{++} concentrations, calcite usually forms when the solution exceeds the $CaCO_3$ solubility product. At high Mg^{++} concentrations, aragonite usually forms. The Mg^{++} concentration in sea water (1330 mg/l) is considerably in excess of the amount needed to prevent formation of calcite, hence inorganic precipitation from sea water most commonly produces aragonite.

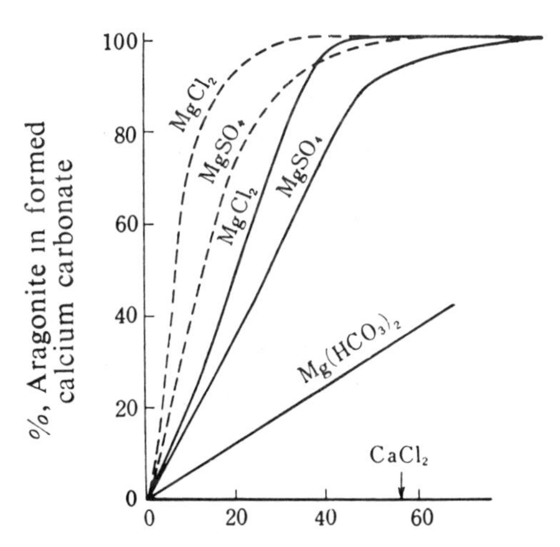

Figure 3–2: The effects of Mg compounds in solution on the mineralogy of precipitated $CaCO_3$. Point on *x*-axis indicated by $CaCl_2$ for precipitation from Mg-free solution. Solid lines-experiments at $10 \pm 2°C$. Dashed lines-experiments at $28 \pm 3°C$. From Kitano, 1962; figure 4.

Simkiss (1964) has suggested that the effect of Mg^{++} is that of a "crystal poison," i.e., the Mg^{++} ions occupy Ca^{++} sites in the growing calcite lattice and block or greatly slow the further growth of the crystal. Due to its small size the Mg^{++} ion is not readily accommodated in the aragonite lattice so that aragonite is able to grow much more rapidly than calcite. Other ions may interfere in a similar manner. Simkiss suggests that the $PO_4^{\equiv}$ ion may act to inhibit calcite formation by occupying the $CO_3^=$ site, and that an organism may control its skeletal mineralogy by controlling the $PO_4^{\equiv}$ concentration at the calcification site. Although this hypothesis has not been adequately tested by chemical determinations at the calcification site, high $PO_4^{\equiv}$ concentrations have been observed in the mantle of growing bivalves (e.g., Bevelander and Benzer, 1948). Similarly, Mg^{++} inhibits apatite formation in natural waters (Martens and Harriss, 1970).

Soluble organic constituents in the solution from which $CaCO_3$ is precipitated also influence the mineral produced (Kitano and Hood, 1965). Kobayashi (1964) showed that the fluid from which precipitation occurs in bivalves with almost wholly calcitic shells contains a single protein fraction while that in aragonitic bivalves contains three fractions. The mechanism by which this control is effected is not known.

Other than Ca, Mg and Sr are by far the most abundant cations in carbonate skeletons. This is related to two factors. The ionic radii and charge of Mg^{++} and Sr^{++} allow them to substitute for Ca^{++} in either the calcite or aragonite lattice. Secondly, both ions are relatively abundant in natural waters. In general, Mg concentration will be greater in calcite, and Sr concentrations greater in aragonite (data summarized by Dodd, 1967). The structural explanation for this relationship is that the small Mg^{++} ion (radius 0.66 Å) substitutes for Ca^{++} (radius 0.99 Å) more readily in the calcite lattice, which is isostructural with magnesite ($MgCO_3$), than it does in the aragonite lattice. Likewise the larger Sr^{++} ion (radius 1.12 Å) substitutes more readily in the aragonite lattice, which is isostructural with strontianite ($SrCO_3$). Mg^{++} can occur in calcite in significant excess of the amount observed by x-ray diffraction, but the phase (or phases) in which it exists is not known (Milliman *et al.*, 1971).

A straightforward physical-chemical model is used to predict oxygen isotope ratios in carbonate skeletons (Epstein, 1959). The equilibration of oxygen isotopes between carbonate and water is generalized to:

$$H_2O^{18} + 1/3\ CO_3^{=16} \rightleftharpoons H_2O^{16} + 1/3\ CO_3^{=18}$$

In order to minimize the free energy of the system, the O^{18} shows a slight tendency to concentrate in the $CO_3^=$ relative to the H_2O. This relationship can be expressed by the equilibrium constant K

$$K = \frac{(H_2O^{16})(CO_3^{=18})^{1/3}}{(H_2O^{18})(CO_3^{=16})^{1/3}} = \frac{[(CO_3^{=18})/(CO_3^{=16})]^{1/3}}{(H_2O^{18})/(H_2O^{16})}$$

where the quantities in parentheses are molecular abundances. For simplicity and by convention the fractionation factor α is usually used instead of the equilibrium constant. The fractionation factor α for the above reaction is defined as

$$\alpha = \frac{(CO_3^{=18})/(CO_3^{=16})}{(H_2O^{18})/(H_2O^{16})}$$

If the O^{18} and O^{16} were evenly distributed, α would equal unity. In fact, at 25°C, $\alpha = 1.021$ (McCrea, 1950), indicating that the O^{18} concentrates in the carbonate relative to the water.

Environmental Factors

Lowenstam (1954a, 1954b) emphasized a pronounced influence of temperature on mineral composition of invertebrate skeletons. He pointed out that this effect is expressed in three ways. (1) Some groups of organisms having aragonitic skeletons are far more abundant in the tropics than in the higher latitudes. The scleractinian corals are the best example of this effect. (2) Certain groups of organisms having aragonitic skeletons are found only in the tropics and semi-tropics. The best examples of this effect are the calcareous green algae. (3) In groups having skeletons composed of a combination of aragonite and calcite, the proportion of aragonite increases as temperature increases. Examples of this effect are found in the mollusks, annelids, coelenterates, and ectoprocts.

Under equilibrium conditions, the Mg/Ca and Sr/Ca ratios in a carbonate skeleton should be proportional to those ratios in the water in which the carbonates formed. The relationships can be expressed as $(M/Ca)_{skeleton} = K(M/Ca)_{water}$, in which M represents the molar concentration of a cation (Mg, Sr, etc.), and Ca the molar concentration of calcium. K is a proportionality constant commonly called a distribution or partition coefficient. The concept of distribution coefficients is useful in studying trace element chemistry of inorganic precipitates. It has been used especially in studying the Sr trace chemistry of carbonates (e.g., Holland *et al.*, 1964; Lerman, 1965a; and Kinsman and Holland, 1969). Odum (1951) varied the Sr/Ca ratio of the water in which the fresh water snail, *Physa*, was growing and showed a direct correlation between the Sr/Ca ratio in the water and that in the snail shell. He even caused the snails to form strontianite ($SrCO_3$) shells by sufficiently elevating the Sr/Ca ratio of the water.

The water chemistry effect is very apparent in comparing the Sr/Ca ratio in mollusk shells from marine and fresh water. Fresh waters have variable Sr/Ca ratios but they are usually low relative to sea water (Odum, 1957a), which is reflected in the generally lower Sr/Ca ratio in fresh water clams and snails as compared to marine forms (Odum, 1957b). In marine skeletons, variations in trace chemistry due to the effect of sea water chemistry should be minor as both Sr/Ca and Mg/Ca ratios are nearly constant (Riley and Tongudai, 1967).

The effect on Mg/Ca ratios in skeletal carbonates of variation in the Mg/Ca ratio of the waters in which they grew has not been studied. The distribution coefficient concept should apply to Mg/Ca ratios in skeletons as it apparently does to the Mg/Ca ratio in inorganic precipitates (Winland, 1969).

The effect of temperature on Mg concentration in calcite is perhaps the best documented case of an environmental factor affecting skeletal chemistry. As first noted by Clarke and Wheeler (1922) and later confirmed by many others (especially Chave, 1954), Mg concentration increases regularly with temperature in many groups of organisms; this is especially apparent in forms characterized by a high Mg content. The effect of temperature on the Sr concentration was discovered considerably later (Pilkey and Hower, 1960) and has been demonstrated for only a few taxa. Pilkey and Hower (1960) observed a negative correlation between temperature and Sr content in echinoids, but Lowenstam (1961) observed a positive correlation between temperature and Sr in articulate brachiopods. Dodd (1965) observed a positive correlation between temperature and Sr in the calcitic outer layer of *Mytilus* but a negative correlation in the aragonitic layer. A positive correlation was noted in the predominantly calcitic bivalve *Crassostrea* by Lerman (1965b) and a negative correlation in the aragonitic bivalve *Cardium* by Hallam and Price (1968).

Urey *et al.* (1951) demonstrated that oxygen isotopic composition of skeletal carbonates varies with temperature. The fractionation factor for the reaction between water and carbonate is constant only at a given temperature. α changes from 1.025 to 1.021 between 0° and 25°C (McCrea, 1950) indicating that at lower temperatures the discrepancy between the O^{18}/O^{16} ratio of the water and carbonate is greater than at higher temperatures, or as the temperature increases the fractionation of O^{18} and O^{16} becomes less and less. The O^{18}/O^{16} ratio of the carbonate should thus vary inversely with temperature provided the O^{18}/O^{16} ratio of the water in which the precipitation is occurring remains constant. Epstein *et al.* (1953) confirmed this by analyzing mollusk shells grown at known temperatures in waters with known O^{18}/O^{16} ratios when equilibrium between shell and water was attained. They were able to derive an empirical temperature equation from these data (*figure 3–3*) which has been used in many subsequent paleotemperature studies.

Epstein and Mayeda (1953) showed that the O^{18}/O^{16} ratio of sea water also varies directly with salinity. If temperature is constant, the O^{18}/O^{16} ratio of shells growing in that water should decrease from normal marine through brackish to fresh water.

The change in O^{18}/O^{16} ratio with changing salinity is explained in the following fashion. H_2O^{16} has a lower vapor pressure and enters into the gas

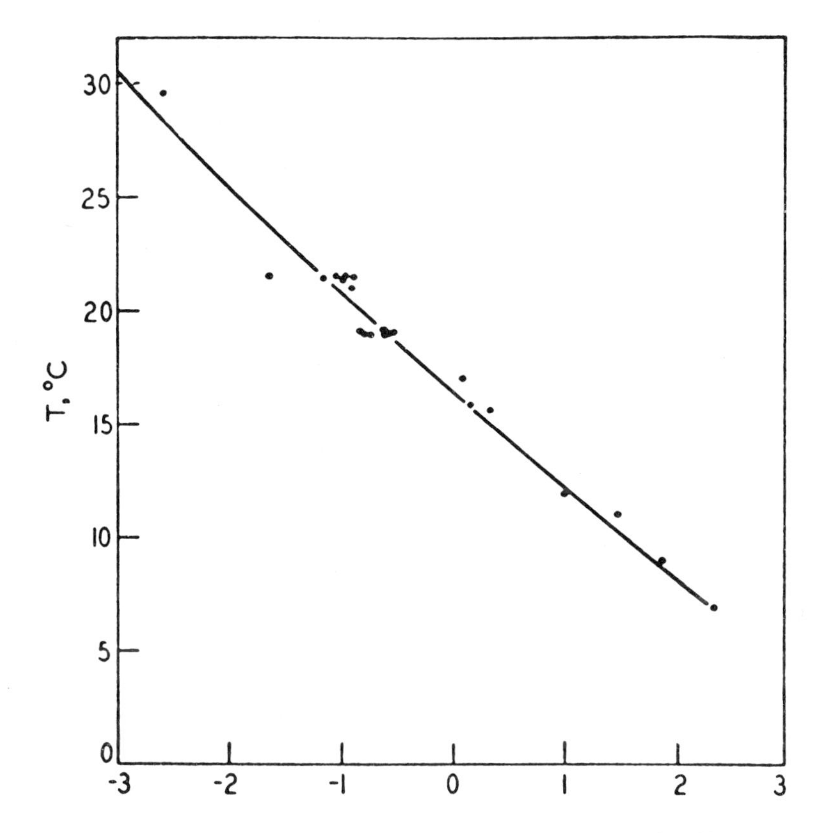

Figure 3–3: Variation with growth temperature of the O^{18}/O^{16} ratio (expressed as $\delta\ O^{18}$ relative to the PDB standard) of molluskan carbonates grown at known temperatures. From Epstein *et al.*, 1953; figure 9.

phase more readily than H_2O^{18}. The fractionation factor for this process at 25°C is

$$\alpha = \frac{(O^{18}/O^{16})_{\text{liquid}}}{(O^{18}/O^{16})_{\text{vapor}}} = 1.008$$

Consequently the water vapor over the oceans initially has an O^{18}/O^{16} ratio about 8‰ lower than that of the ocean itself. When the vapor first starts to condense and return to the liquid phase as rain, H_2O^{18} will tend to condense first as it is favored in the liquid phase. The initial rain will have an O^{18}/O^{16} ratio, which is the same as the ocean water, but the vapor which remains will have a still lower O^{18}/O^{16} ratio due to the loss of O^{18}. Subsequent rain will always have an O^{18}/O^{16} ratio 8‰ higher than the vapor, but the vapor will get lighter and lighter as more O^{18} is removed in the form of rain (Epstein, 1959). Rain falling on the continents usually has an O^{18}/O^{16} ratio several per mille lighter than that in sea water. The exact composition of the water will depend on the condensational history of the water vapor from which it was derived. In general the lowest O^{18}/O^{16} ratios are produced at the coldest temperatures after most of the vapor has been condensed and only the last, lightest portion remains. This lightest precipitation falls in the polar regions and may be as much as 50‰ lighter than sea water.

In a few exceptional cases the O^{18}/O^{16} ratio of fresh water may be higher than that in normal sea water. If a fresh water body is subject to intense evaporation, it will lose a great deal of H_2O^{16}. The remaining water will then be enriched in H_2O^{18} due to the loss of the H_2O^{16}. Such a condition prevails in the fresh water of the Everglades of Florida during the dry season (Lloyd, 1964). This same enrichment process can of course also occur in enclosed or semi-enclosed bodies of marine water.

Genetic Factors

On physiochemical grounds, skeletal carbonates should only exist as calcite. Yet $CaCO_3$ skeletons may occur as either calcite or aragonite or as a combination of the two, and rarely even as vaterite. An additional assortment of non-carbonate minerals also occurs in invertebrate skeletons (reviewed by Horowitz and Potter, 1971, Table 8). Clearly there is a strong genetic control over skeletal mineralogy, possibly primarily influenced by the effect of the organic substrate on which the precipitation occurs (Glimcher, 1960; Wilbur and Watabe, 1963).

Many groups of organisms are characterized by specific ranges of concentrations of major and minor elements, and by the degree of oxygen and carbon isotope fractionation. These differences are produced by differences in the biochemical processes of skeletal formation. Although the Mg and Sr concentrations in skeletal carbonates of various organisms show no unequivocal trends, a generalized pattern towards lower Mg and Sr concentrations in more highly organized groups has been suggested (Chave, 1954; Lowenstam, 1963). Mg in aragonite shows this trend (*figure 3–4*), which is more recognizable in this example than many because an independent temperature influence was negated by using organisms from a single locality (Lowenstam, 1963). This was a very useful "natural experiment."

Although skeletal calcites tend to occur in two groups on the basis of their Mg content (low Mg-calcite with less than 5–6 mol per cent $MgCO_3$, and a high Mg-calcite group with more), they do not rigorously reflect the order originally predicted (Chave, 1954) from increasing phylogenetic position. The lowly planktonic foraminifera and coccoliths have much less Mg than is predicted, and the echinoderms much more. For calcites precipitated from normal sea water at 25°C, the Sr/Ca ratio should be about 1.3×10^{-3}

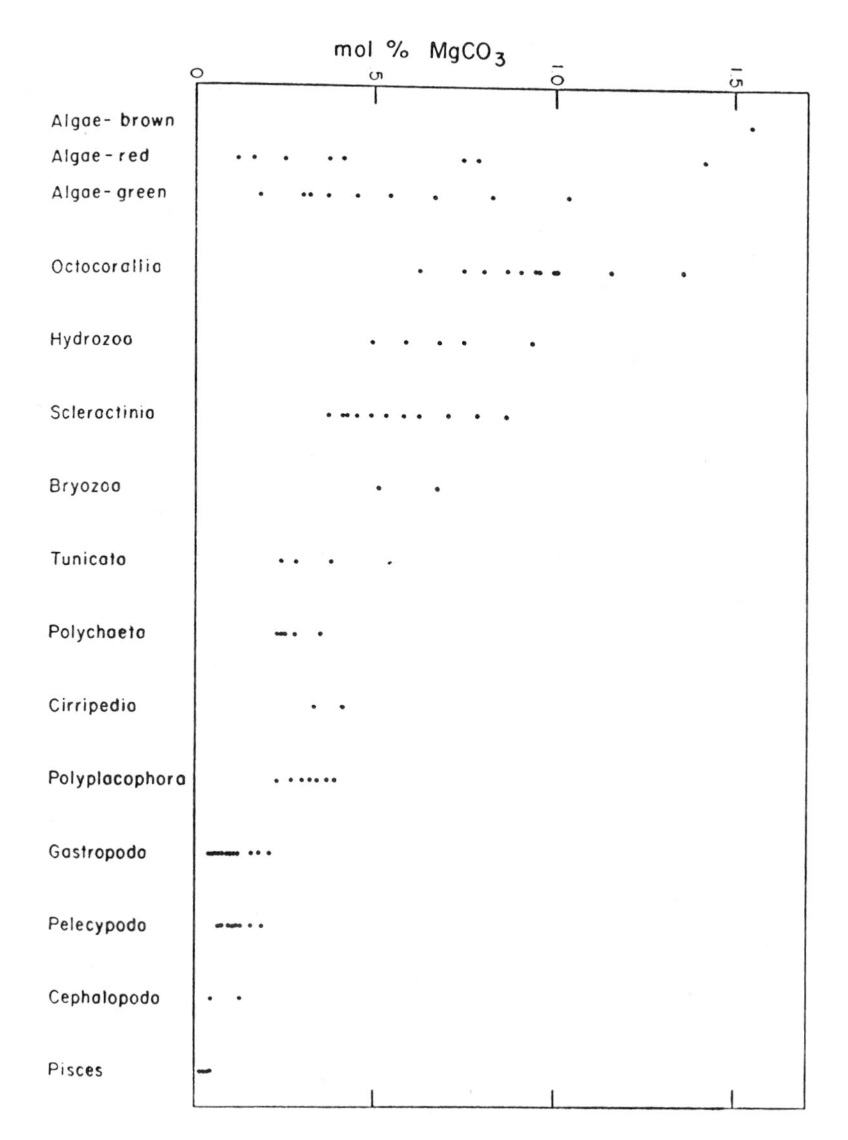

Figure 3–4: Mol % $MgCO_3$ in aragonitic skeletons of various taxa arranged in order of increasing organic complexity. All samples from Bermuda. From Lowenstam, 1963; figure 8.

(Holland *et al.*, 1964). Molluscan calcites have about this value, but most other skeletal calcites have a ratio of Sr/Ca about twice this amount.

Skeletal aragonites also tend to occur in high and low Sr groups. The high Sr group (Sr/Ca ratio of about 8–11 × 10^{-3}) is typified by corals and green algae. Their ratio is similar to that in aragonite precipitated inorganically from sea water, and is slighly above the Sr/Ca ratio in sea water. The low Sr group (Sr/Ca ratio of $1–3 \times 10^{-3}$ in most instances) includes the molluscs (excepting chitins) and fish with their aragonitic otoliths. Exceptions to the trend of decreasing Sr with higher phyletic level in skeletal aragonites include annelids and barnacles.

The general trend of increasing physiological control over the composition of mineralized tissues the higher the phylogenetic position may be mirrored by long-term trends within a single phylum. Evidence was cited for changes in degree of fractionation of oxygen and carbon isotopes since the Devonian for echinoderms (Weber and Raup, 1968). And the progressive decrease in Sr/Ca ratio of Pennsylvanian, Cretaceous, and Recent snails is attributed to improved physiological control of mineralization (Lowenstam, 1964).

In addition to these long-term phylogenetic influences, some bivalves and echinoderms show changes in composition during their life span. The Sr/Ca ratio in the calcitic outer shell layer of the bivalve genus *Mytilus* decreases during ontogeny (Dodd, 1965) and some genera of fresh water bivalves at an age of 5–10 years abruptly increase the Sr/Ca ratio of material being deposited (Nelson, 1964). Oxygen and carbon isotope ratios change as echinoderm spines elongate (*figure 3–5*; Weber and Raup. 1966a).

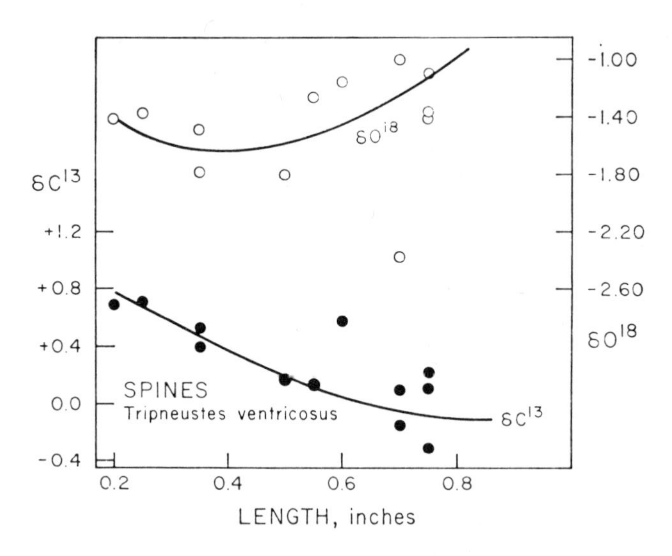

Figure 3–5: Isotopic composition of spines of the sea urchin *Tripneustes ventricosus* versus length of the spine. From Weber and Raup, 1966a; figure 5.

In cheilostome ectoprocts, a series of species collected in the Woods Hole area had different amounts of Mg in their calcite, but the variation did not make sense in terms of local environmental situations. These variations follow the general pattern of lowest Mg in species with most northerly ranges and highest Mg in species with most southerly ranges (Schopf and Manheim, 1967, fig. 6). Considering the difficulties in plotting accurate ranges, one sees that these data approach a cline in Mg according to species distribution, as though natural selection had acted so that species now maintain an amount of Mg in their calcite appropriate for the range of temperatures over the distribution of the *species* (not the specimen).

The degree of fractionation of both oxygen and carbon isotopes in echinoderms is dependent upon the species, genus, family, suborder, or order studied, and, to a lesser but noticeable extent, the part of the skeleton examined (spine, lantern, position of plate, etc.) (Weber and Raup, 1966a, 1966b; 1968; Weber, 1968).

The explanation for different species or higher taxa having characteristic chemical compositions can be given on more than one level. At the physiological or biochemical level, the question becomes "How does species A discriminate against Mg, or how does species B concentrate Mg?" This is the approach paleontologists seem to talk about, and in which they take an active interest. Comparative studies of the process of mineralization are an exceedingly active field at the moment and more answers at the physiological level may soon be forthcoming (see Watabe and Wilbur, 1960; Hare, 1963; Wilbur, 1964; Degens, Johanneson, and Meyer, 1967). On the other hand, at the population level, the question becomes "Why has natural selection resulted in species A discriminating against Mg?" At present most studies treat Mg (or Sr, etc.) like a black box and do not consider its long-term biological significance.

Diagenetic Factors

Diagenesis is often very important in determining the final composition of a fossil. The general effect of diagenesis is to obliterate the "signal" imparted to the fossil by the physiochemical, genetic, and environmental factors. The purpose of this paper is mainly to point out the "signal" rather than the "noise," so we will say little on diagenesis here. A consideration of the possible effect of diagenesis is of course vital to any study of fossil chemistry and is one of the most important factors limiting the usefulness of biogeochemical studies. In very general terms, the older the fossil, the more likely it has been affected by diagenesis. Unaltered Cenozoic fossils may be quite common, but unaltered Paleozoic fossils are rare. More recent studies have considered and attempted to evaluate diagenesis (e.g. Stanton and Dodd, 1970). In biogeochemical studies as indeed in all studies of a historical nature, the best evidence against an altered record is that independent lines of evidence do not disprove the hypothesis being tested.

Conclusion

Present survey work, far from consisting of a carefully chosen suite of specimens from specific environments, usually includes a few specimens that happened to be at hand. Thorough analyses investigating individual variation, species variation, and environmental variation are few. Many specific questions relating composition to biological variables, such as the uniform deep-sea environment vs. the variable subtidal environment, remain to be investigated. Especially in need of being carried out are studies that explain the process of mineralization; hopefully these results will, on the one hand, lead to comprehensive models of chemical composition independent of particular species and, on the other, be carried out with a historical view of the process so that an explanation in terms of natural selection is also forthcoming.

PART III

Population and Evolution

4

MODELS INVOLVING POPULATION DYNAMICS

Anthony Hallam

Editorial introduction. The study of populations is concerned with both the average and the cumulative effects of individuals. This field of study has come into its own as a distinct entity in the 1960's (see Connell, Mertz, and Murdoch, 1970) and may have an important bearing on paleontology in the 1970's. To date, the chief paleontologic application is the significance of size-frequency analyses, as indicated by Hallam. However, as Hallam goes on to state, the invention of the concept of *r*-selection (for reproductive ability) versus *K*-selection (for resource utilization), and the realization that life tables can be constructed in favorable fossil material, serve to provide a new focus for this important level of organization.

The genetic differentiation of populations is equal in importance to demographic characteristics of species. This differentiation can now be mapped on a locus by locus basis over the geographic range of natural populations. Plots that show the way in which gene pools are spatially (and temporally) organized will form a factual basis for paleontologic speculation about the coherence of gene pools. Maps showing genetic differentiation over distances from square meters to populations separated by 1500 km are beginning to be prepared for marine forms (Gooch and Schopf, 1970; Schopf and Gooch, 1971). This topic is not further considered in the book because the information that is of direct paleobiologic interest is still fragmentary.

Introduction

The study of animal and plant populations as functioning systems, involving the processes that control rates of growth, loss and replacement, is a flourishing field of ecology with many economic applications. That population dynamics has any bearing on the study of fossils is by no means self-evident, but a proportion of the information obtainable from fossil assemblages can be analyzed in these terms to produce insights unobtainable in other ways. Not only can valuable clues emerge about events during the life of the organisms in question, but a full understanding of inorganic processes affecting organisms between death and fossilization demands at least an elementary knowledge of population dynamics.

General accounts of population dynamics in modern ecology are available in the works of Deevey (1947), Slobodkin (1961) and Solomon (1969). Particularly useful are the superb discussions accompanying the readings assembled by Connell, Mertz, and Murdoch (1970), and the self-teaching *Primer of Population Biology* (Wilson and Bossert, 1971). More advanced mathematical treatments of the subject as related to fisheries are given by Beverton and Holt (1957) and Nikol'skiĭ (1969). The application to fossils, notably the Pleistocene cave bear, was reviewed to the early 1960's by Kurtén (1964). Most paleontological work has concerned itself, however, with invertebrates, notably the Mollusca, and this is reflected in the examples chosen in this chapter.

In the ensuing account, consideration will be given initially to the dynamics of populations as they are observed at the present day, and a series of models developed. Then the application to fossils will be discussed. It is convenient to separate *short-period changes*, of the order of months to at most a few years, from *long-period changes*, of the order of years to decades or even longer.

Short-period Changes

The basic data of population dynamics relate to the age of organisms and their rates of growth, mortality and recruitment.

Growth rates. The age of some animals and plants is determinable by concentric growth rings on their hard parts, such as bivalve shells, fish scales, the genital plates of echinoids and tree rings, or by degrees of relative wear, for instance in mammalian teeth. Useful references to this subject include Roughton (1962), Runcorn (1966), Neville (1967) and Klevezal' and Kleinenberg (1969). Particularly useful are the annual rings formed on the shells of many Bivalvia because of a slowing down or cessation of growth in winter. The possible ambiguity as to whether a given ring expresses a seasonal retardation of growth or merely records a short-lived disturbance can be resolved by microscopic study, which may in favorable cases reveal diurnal layers (Pannella and MacClintock, 1968).

Growth ring analysis shows that bivalves generally undergo a gradual, more or less exponential decline of growth rate with age (*figure 4-1*). Unlike that of mammals, growth does not cease with the attainment of maturity but merely slows down. Such a pattern appears to be characteristic of many invertebrates. The establishment of the relationship of size to age is obviously a matter of importance because many bivalves and other invertebrates do not possess clearly defined annual rings. Levinton and Bambach (1970), from an analysis of bivalve growth data presented by Hallam (1967), concluded that the following equation gives a good general approximation of bivalve growth.

$$D = s \ln (T + 1)$$

where D = size, T = time, and s is a constant.

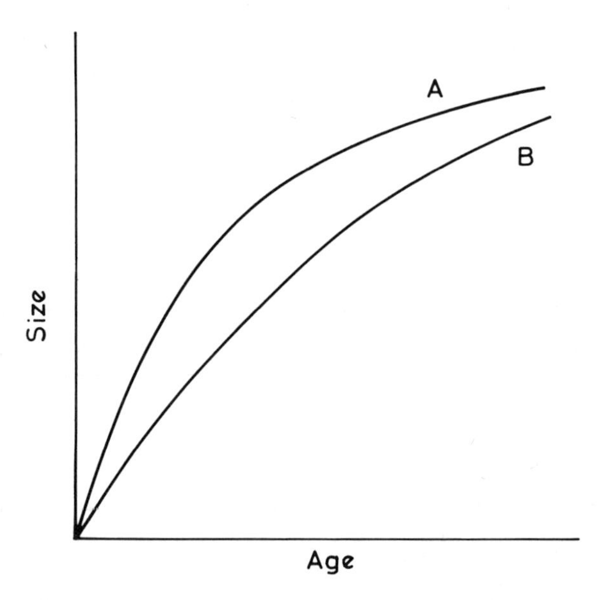

Figure 4–1: Schematic and simplified growth curve for bivalves. A = high initial growth rate, B = low initial growth rate.

In detail the growth is probably more complex, since a simple exponential function fails to account for the sigmoid growth curves that are often observed, with the point of inflection near the origin, and for the growth of the oldest bivalves, which may be greater than calculated; it appears that the exponent may often itself decrease exponentially with time (Hallam, 1967). (Quite possibly the sigmoid shape is due to a low initial growth rate of bivalves settled as spat in the fall.) However, the Levinton-Bambach equation should prove useful for many purposes.

The relationship of growth rates to latitude should be noted briefly. It appears, notably from the work of Weymouth and Rich (1931) on the Pacific razor clam *Siliqua patula*, that in low latitudes a relatively high growth rate is combined with relatively early death. In high latitudes slower growth correlates with lower temperatures and hence lower metabolic rates, but longevity and hence the ultimate size attained are greater. Intermediate latitudes show correspondingly intermediate growth and mortality characteristics (*figure 4-2*). Where growth ring analysis is feasible there is a potential application to paleoecology for times in the past when latitudinal temperature gradients were appreciable.

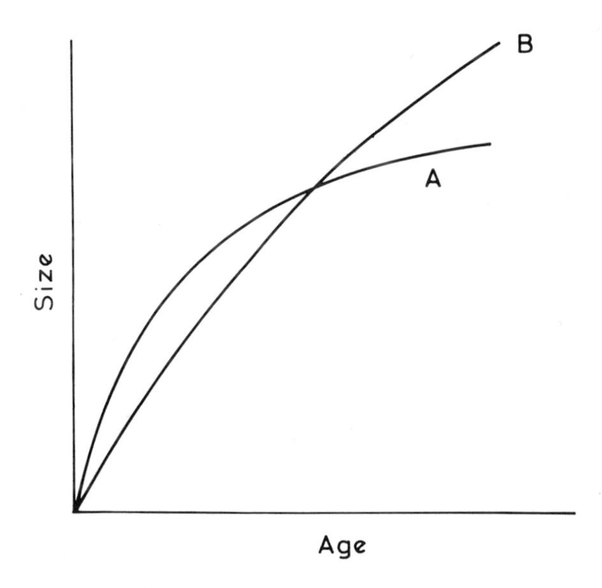

Figure 4–2: Possible variation of intraspecific growth rate with latitude. A = low latitude (growth rate faster but longevity lower); B = high latitude (greater maximum size attained).

Mortality rates. It is important to distinguish two types of mortality. The first, or *census*, type, is catastrophic, all age classes being wiped out suddenly, so that the newly created death assemblage will be identical in age and size characteristics to the life assemblage from which it was produced. The second, or *normal*, type relates to mortality caused by disease, old age, spasmodic accidents and predation, and is of the sort one might encounter in a human graveyard, at least in times of peace and plenty. These different types of mortality can be subjected to different types of analysis, respectively termed Time-Specific and Dynamic (Kurtén 1964), but the distinction is quite subtle in practice and can be disregarded here.

The rate of mortality is usually expressed as the percentage of the population dying each year, and the complete dynamic data are presented in *survivorship curves* and *life tables*. The latter give estimates of life expectation at different ages and are of more concern to actuaries than paleontologists; methods of calculation are given by Deevey (1947).

In a survivorship curve, the number of survivors from 1000 individuals born at the same time is plotted along the ordinate on a logarithmic scale against the age at death along the abscissa. The age can be given either in absolute or in relative terms, in which latter case it is expressed as percentage deviation from the mean length of life. This measure allows meaningful comparisons to be made between organisms of widely differing longevity.

Plotting the points of a survivorship curve is done in the following manner. Take for example a sample of 1000 individuals. If 300 die within the first year of growth, then the number of survivors at the end of Year 1 is 700. If 250 die within the next year, then Year 2 plots as 450; with a further 200 deaths in the following year, the number of survivors at the end of Year 3 is 250; and so on until the whole population is dead. Plotting numbers on a logarithmic scale has the advantage that at a constant rate of mortality the survivorship curve plots as a straight line of negative slope.

The slope of the survivorship curve is proportional to mortality rate, and variations of steepness along the length can be used to distinguish mortality patterns (*figure 4–3*). Curves convex towards the upper right signify mortality rate increasing with age; concave curves signify the reverse. The former is characteristic of a number of higher vertebrates, with our own species showing the most convexity. Improved nutrition and medical care have served to reduce mortality in the earlier years to very low rates by the standards of previous generations or the animal kingdom at large, whilst absolute longevity has not been affected. The result is a highly convex survivorship curve with a near-horizontal portion changing abruptly to a near-vertical one.

The mortality rate in the larval stages of invertebrates must be enormous and the survivorship curves are in consequence strongly concave. However, mortality after the settlement of spat on the sea bed is another matter and much more relevant to the interests of the paleontologist. Information on post-larval mortality is much less than that on growth, and such data as there are, on bivalves for instance (Hallam, 1967), indicate a wide range of variation. It appears to be the rule that mortality in the first year is higher than in subsequent years. Perhaps the most normal type of survivorship curve for post-larval bivalves and other types of benthonic invertebrate is sigmoidal, with high juvenile mortality being followed by a decline in youth and maturity, to increase again in old age (Cadée, 1968).

Recruitment rates and population growth. To counter mortality, populations must be renewable by the birth of new individuals or the influx of organisms from elsewhere, for instance in the benthos as settling larvae or as adult migrants. Alternatively, during colonizing episodes populations may develop where nothing existed previously. A great deal of attention has been paid to this phenomenon (see, for instance, Cole, 1954; Lewontin, 1965; and MacArthur and Wilson, 1967).

Understanding basic theory in this field requires calculus. The following brief discussion indicates some main elements and uses the terminology of Connell, Mertz and Murdoch (1970), which the interested reader should consult.

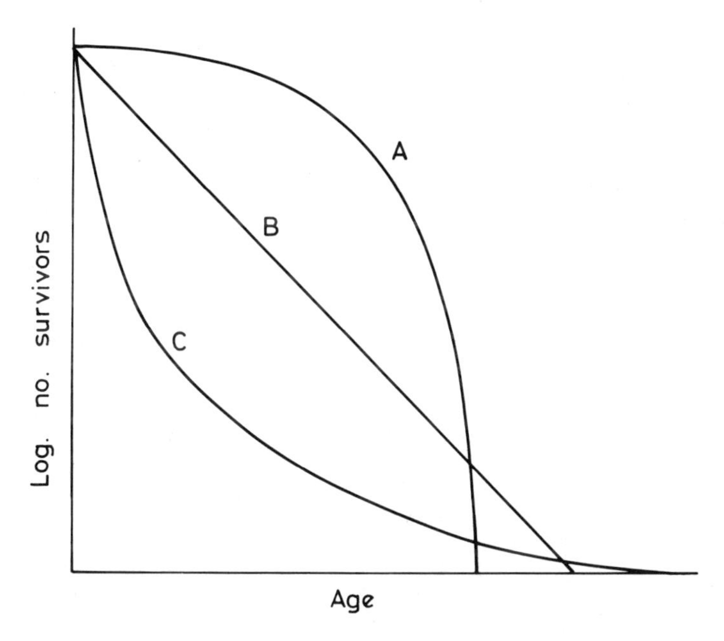

Figure 4–3: Variation in survivorship curves. A = mortality increasing with age, B = mortality constant, C = mortality decreasing with age.

We assume that at the initial time ($t = 0$), the growth of the population (N) is proportional to the size of the initial population (N_0). If a stable age distribution is achieved in the population, the rate of change of the population (dN/dt) is equal to a constant r (called the ultimate rate of natural increase by some, and the intrinsic rate of increase of the population by others) times the original population at a given time (N_t). Or, $dN/dt = rN_t$, which on integration gives $N_t = N_0e^{rt}$. Note that if $r < 0$, then the population decreases; if $r = 0$, the population stays the same; and if $r > 0$, the population increases.

Let K be the number of organisms a habitat can hold at saturation; then $K - N$ is the degree of unsaturation. As population size increases, resistance to further growth increases, as saturation is approached. Graphically the change in size of population with time can be expressed by the familiar sigmoid "logistic growth curve" (*figure 4–4*). This shows how the rate of increase diminishes progressively as the saturation point for the environment is reached (e.g., by crowding or limitation of food). This may be expressed by the equation

$$\frac{dN}{dt} = \frac{r(K - N)N}{K}$$

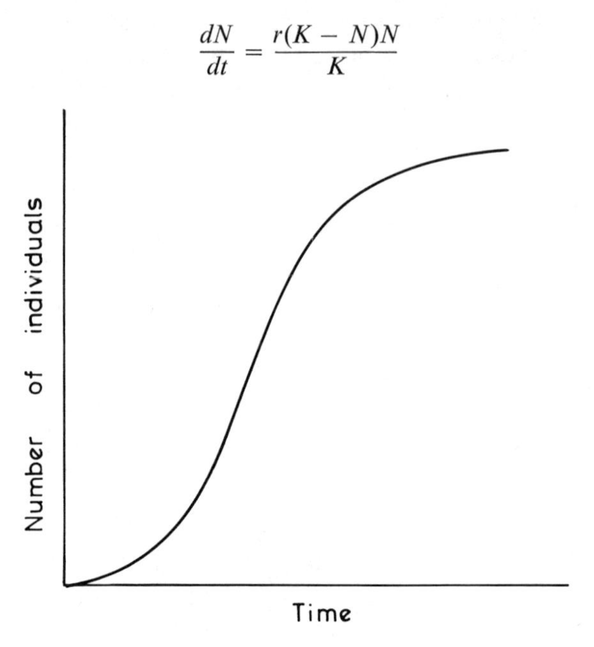

Figure 4–4: The logistic growth curve.

For populations, l_x is the probability of living from birth to age x, or the age-specific survivorship; m_x is the expected number of daughter births which a female will produce at age x if she lives to age x, or the age-specific birth rate. The product $l_x m_x$ summed over all ages yields the expected number of daughters produced by females who enter the population and is called the Net Reproductive Rate, R_0. A widely used expression relating these to r in natural populations is called the Volterra equation, which holds that

$$1 = \sum_{0}^{\infty} e^{-rx} l_x m_x \quad \text{or} \quad 1 = \int_{0}^{\infty} e^{-rx} l_x m_x \, dx$$

These methods of analyzing natural populations have so far been assimilated into few paleontological problems. Predictions from theory indicate that shifting reproduction to younger ages increases r far more than does an increase in fertility; in fact "small absolute changes in developmental rates of the order of 10% are roughly equivalent to large increases in fertility on the order of 100%" (Lewontin, 1965). Changes of this degree ought to be able to be determined in appropriate paleontological collections where the age of reproduction can be determined in successive populations. Indications that this and similar approaches will be increasingly utilized are given by Levinton and Bambach (1970) for bivalves and Kaufmann (1970, 1971) for ectoprocts.

Over a wide range of latitude, from the margins of the subtropics almost to the poles, breeding and recruitment of invertebrate spat from the larvae is confined to a few weeks in the spring and fall, though often one of these seasons in a given place is much more important than the other. In low latitudes recruitment might be spread over a larger part of the year as the seasonal climatic effect is diminished, but pertinent data are sparse. From year to year the amount of successfully settled spat may vary considerably (e.g., Hancock and Simpson, 1961). Furthermore, different recruitment schedules may be selected for in different environments (*r* and *K* selection), as will be discussed later.

Construction of Size-Frequency Models

For an assemblage of dead shells accumulated from a living population through normal processes of mortality, modal analysis of growth rings (Craig and Hallam, 1963) will give the necessary age data for the construction of a survivorship curve. Where growth rings are not clearly discernible, a "size-survivorship curve" can be constructed from the size-frequency histogram as indicated in *figure 4–5*. To convert this to a proper survivorship curve a realistic model of age-size relationship, such as that mentioned earlier, must be used, and the appropriate figures read off. Such a method can of course give only approximate results but this will usually be sufficient for paleontologists.

A wide variety of size-frequency models for invertebrates can be generated by realistically varying growth, mortality and recruitment rates. A preliminary attempt at this (Craig and Hallam, 1963) was followed by a more comprehensive analysis using a computer (Craig and Oertel, 1966). In this important work a large number of histograms was generated; these set limits and offer simplified patterns of what might be expected in the natural state. Attention will be confined to a few generalized size-frequency models involving both living populations and death assemblages derived therefrom.

Living populations. If recruitment is confined to certain limited periods of the year, as seems to be the case in most environmental regimes, then a bimodal or multimodal size-frequency histogram will be recorded, because the time gap between spat settlements (or other types of juvenile recruit) will allow growth of the older generation. Each new brood will tend to be distributed symmetrically about a mode in the usual way, and the variance will

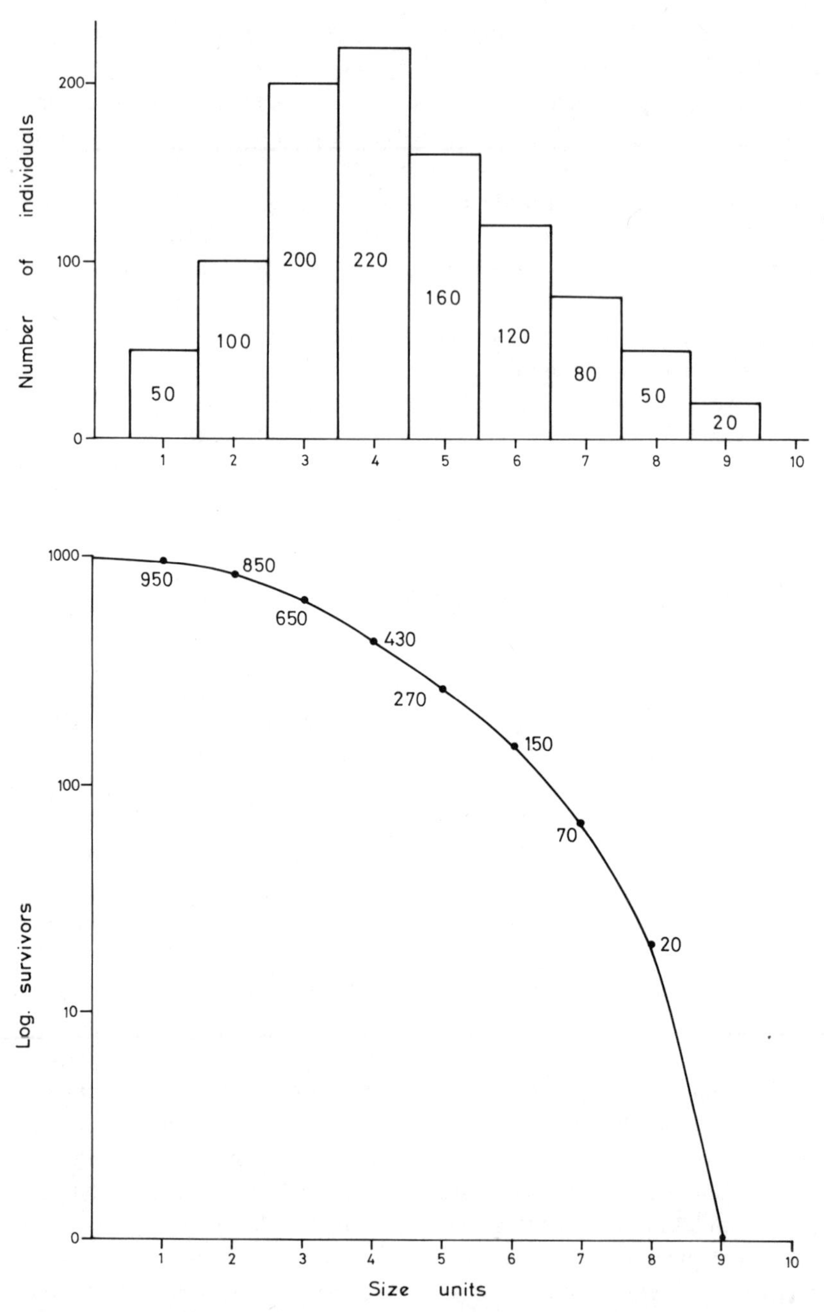

Figure 4–5: Diagram to illustrate how a "size-survivorship" curve may be constructed from a size-frequency histogram. Numbers in a histogram refer to individuals in each size class; numbers in curve signify survivors at a given size.

increase with growth. As growth slows down progressively with the onset of maturity after a year or so the successively generated peaks will merge so that size-frequency distributions will rarely show more than two or three modes (*figure 4–6*). (It is always desirable to measure samples of at least hundreds of specimens to eliminate additional spurious peaks and troughs.) Examples of multimodal size-frequency histograms that can be related to successive annual broods of benthonic bivalves are given by Craig and Hallam (1963) and Schmidt and Warme (1969).

So far no allowance has been made for longer-term variations of recruitment, as for instance when a "lean year" for settlement follows a good one. Such a circumstance can, of course, also generate separate peaks, as argued specifically for certain New Zealand brachiopods by Rudwick (1962). The only way the two types of distribution can be distinguished is by obtaining independent evidence of the age of the specimens, as from growth ring analysis.

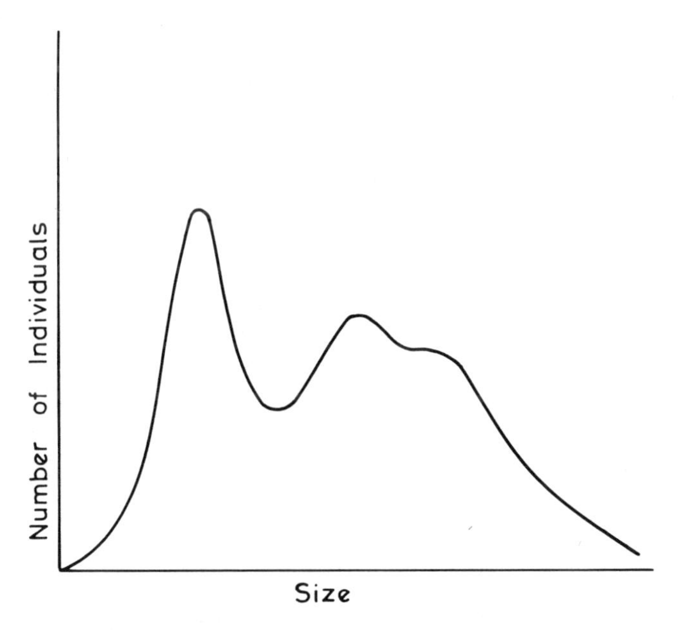

Figure 4–6: Bimodal size-frequency curve of a life assemblage of several years' seasonally-varied growth.

Death assemblages. As a general rule assemblages of dead shells derived directly from living populations tend to give rather smooth unimodal size-frequency distributions, always provided of course that the samples are suitably large in size. The shape of such distributions is dependent substantially on the growth and mortality rates. Consider the size-frequency models of *figure 4–7*, which embrace the range of variation encountered in nature, from the strongly positive skew distribution (case 1) to the negative skew distribution (4) via the moderately positive skew (2) and the normal distribution (3). Case 1 is produced by a combination of high juvenile mortality rate subsequently decreasing with maturity, and low initial growth rate. The distribution will shift progressively toward case 4 with a change toward high initial growth rate and mortality rate increasing with age. It is impossible to determine the relative importance of the two factors without independent evidence of growth rate.

Polymodality in death assemblage size-frequency distributions, though apparently exceptional, can occur if, say, mortality increase coincides with a seasonal cessation of growth, as often happens during the winter in cool temperate regimes. This will only occur, of course, if polymodality is also present in the life assemblage from which it is derived due to restriction of recruitment to a short period of the year.

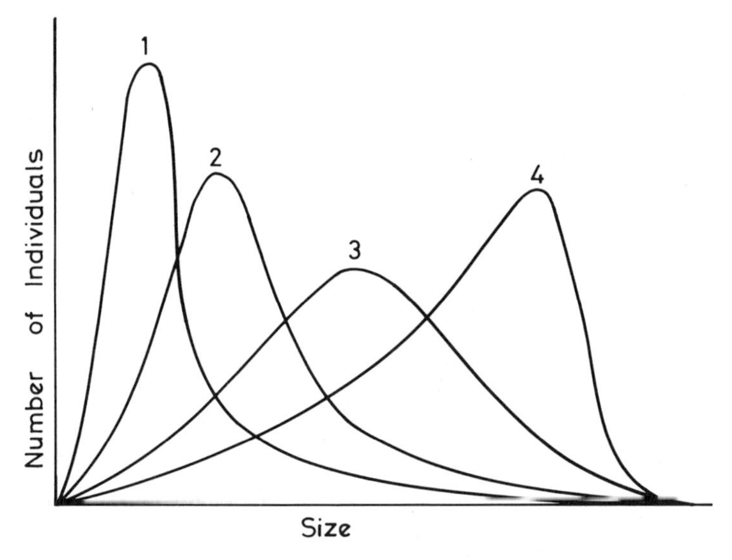

Figure 4–7: Four size-frequency model distributions embracing most natural conditions. 1 = strongly positive skew, 2 = moderately positive skew, 3 = normal, 4 = negative skew.

With regard to the relative importance in actuality for settled benthos of the four model distributions portrayed in *figure 4–7*, the moderately positive skew is probably rather more common than the strongly positive skew, the normal is unusual and the negative skew highly exceptional (e.g., Craig and Hallam, 1963; Craig, 1967; Levinton and Bambach, 1970; Schmidt and Warme, 1969). Even without growth data it must be concluded that mortality tends in general to decline from youth to maturity, though it may increase again at a late stage. Dynamic analysis should bring this out by revealing sigmoid shapes for the survivorship curves.

Long-period Changes

What are known as population fluctuations are widespread in the animal kingdom, the numbers of species drastically increasing and decreasing over periods ranging from a few years to several decades. The year-to-year variations in recruitment of benthonic invertebrate populations, referred to earlier, are in effect but one short-period category of fluctuation. Whereas a more or less regular pattern of fluctuation has been claimed for certain mammals, no evidence exists of any such cyclicity among invertebrates (Coe, 1957). There is no evident correlation moreover between different species.

Though all invertebrate species are liable to fluctuations in numbers over the years, there is great variability in capacity to fluctuate. Certain species may increase a thousandfold or more over a period of several years, thereafter to decline just as dramatically. Coe (1957) has termed these *resurgent populations*. Decline in benthonic populations is often catastrophic and may be due to a variety of factors, such as harsh climatic conditions, drastic alteration of salinity (in marginal waters), disease, starvation due to crowding, dinoflagellate blooms or changes in the ocean current systems that bring in the larval recruits. Significant population expansion may be more difficult to explain, except in those well-known cases where Man was responsible for changing the physical environment or introducing exotic species.

Not much has been written on the variability of the capacity of species to fluctuate in numbers. It has been suggested that the extent of pelagic existence of the larvae may be of major importance, since the plankton is susceptible to a number of vicissitudes which would go unrecorded at the site of spat settlement, and there is indeed some evidence that bivalve species without pelagic larvae may be more stable in numbers (Moore, 1958, p. 317). MacArthur (1960) has drawn a distinction between *equilibrium* and *opportunistic* species, the latter being capable of rapid expansion in population because of high fecundity and short generation time. Such species might appear in large numbers for a few years in areas normally inhabited by species which have, as it were, attained an equilibrium with the environment and remain fairly steady in numbers over the years. The opportunistic species are in no such equilibrium, and though they may dominate the fauna numerically for a short time, they are liable to disappear as dramatically as they arrived. A simple model to illustrate the contrast between the two types of organism is presented in *figure 4–8*.

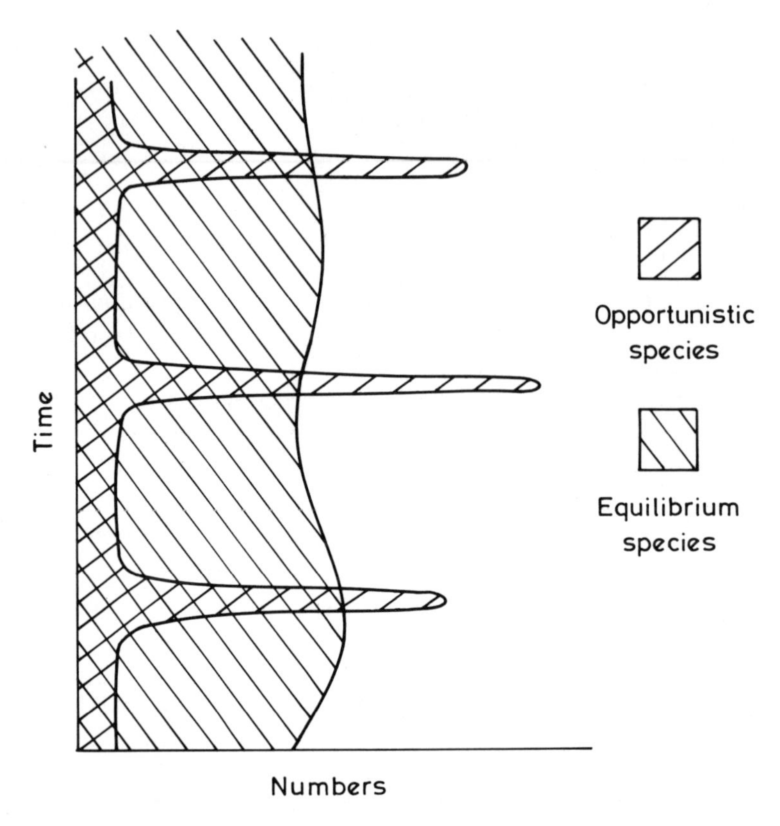

Figure 4–8: Suggested variation in abundance patterns of opportunistic and equilibrium species in a stratigraphic succession.

MacArthur and Wilson (1967) have pursued this matter further in terms of evolutionary biology by introducing the terms *r selection* and *K selection*. The former signifies selection favoring a higher population growth rate and productivity. It is likely to come to the fore during colonizing episodes, so that in environments with no crowding, those phenotypes which harvest the most food (albeit wastefully) will exhibit the highest fecundity and hence be the most fit. The same applies to environments of strong, irregular contrasts, or low predictability, such as coastal waters in high latitudes (Sanders, 1968). Such environments will, in other words, favor opportunistic species.

Conversely, *K* selection favors the more efficient utilization of resources, so that the fittest organisms are those that can replace themselves successfully at the lowest food level. This type of selection may be expected to be characteristic of stable populations in more or less uniformly benign environments of high predictability, such as the deep ocean or the sublittoral zone in the tropics (Sanders, 1968). Equilibrium species will be favored in environments of this type. Perhaps one should add that eurytopic species might be expected to have differing birth and mortality schedules depending on whether their environment in a given instance is predictable or not.

Application to Fossils

Most shell beds as we observe them in the rocks must have accumulated over many years. Bearing in mind the work of Johnson (1965) and others in the modern littoral zone, which reveals that there is often little quantitative relationship between the size-frequency distributions of living populations and accumulations of dead shells of the same species, the possibilities of applying in a straightforward way the concepts of population dynamics to paleontology seem somewhat unpromising. However, the situation is more hopeful than may appear at first sight.

Life assemblages. In the first place, one may be lucky enough on occasion to find assemblages of fossils representing populations which have undergone catastrophic death and thus signifying life assemblages subjected, as it were, to a census. A good example is provided by the substantially monospecific clusters or "nests" of several types of brachiopod in the Marlstone Rock bed (Lower Jurassic) of the English Midlands. These clusters contain unworn articulated specimens of a wide size range and cannot have been swept together by currents into depressions on the sea floor because they do not match the overall faunal composition of the bed. There are plenty of modern analogues, moreover, to confirm that the clusters were original colonial assemblages, which were probably overwhelmed after several years' growth by shifting sediment. Size-frequency analysis of carefully collected material clearly reveals polymodality in some instances. The different peaks have been interpreted as successive broods (Hallam, 1961) but Rudwick's (1962) observations point out that it is possible that the prime cause was varying recruitment levels in successive years. In this connection Sheldon (1965) may be overoptimistic in hoping to estimate growth rates from measurement of peak-to-peak distance in polymodal size-frequency distributions in fossil assemblages, without an independent check such as growth rings.

Former life assemblages whose size-frequency structure has been "frozen" *in situ* might be recognized by several criteria. They should be unworn and, if bivalved, substantially articulated; they should occur in lenses or thin beds dominated by one or two species and not be a random sample of the whole fauna of the beds; if seasonal growth rings are present an estimate of the time of death with respect to these rings may be possible. It may thereby be approximately determinable whether or not a large number of organisms died simultaneously (Rhoads and Pannella, 1970).

Death assemblages. Even in those much more abundant cases where only a death assemblage is present, and in which the size-frequency distribution is the resultant of many years' growth, a knowledge of population dynamics can throw light on a variety of problems. Nowhere is this more clearly evident than in the understanding of the processes which affect shells between death and fossilization.

Many paleontologists have been impressed by the contrast between the observed and expected size-frequency distributions of fossil assemblages (e.g., Boucot, 1953; Olson, 1957). Whereas, given the known high juvenile mortality of invertebrates, the size-frequency distribution should be strongly positively skewed, a large proportion of fossil assemblages appear to have roughly symmetrical distributions approaching the normal. A number of factors have been invoked in explanation.

For a minority of invertebrates whose adult and post-larval juvenile habitats differ, migration of living organisms is a possibility. This was proposed, for instance, for species of *Nassarius* and *Chlamys* in a coastal bay of northern Spain (Cadée, 1968). Since some living cephalopods have inshore breeding grounds the adults tend to be found in more offshore habitats, and this has led Ziegler (1962) to suggest something similar for certain Upper Jurassic ammonite species, whose size-frequency distributions vary from rich to poor in juveniles in different localities. In the case of ammonites, however, especial care is needed to distinguish between juveniles and micromorphic adults. Callomon (1963) claims that macroconch: microconch ratios for given ammonite species in the same bed may differ by as much as 100:1, implying sexual segregation. The only thorough size-frequency analysis of ammonites with sexual dimorphism in mind is that of Lehman (1966), who studied a species of *Eleganticeras* from the Lower Jurassic of Germany. His plotted size-frequency distribution shows a moderately positive skewed curve with two subordinate peaks at sizes higher than the mode. Of 1809 specimens measured, 1223 were attributed to juveniles (responsible for the modal peak), 217 to microconch adults (the intermediate peak) and 367 to macroconch adults (the remaining peak).

Sexual dimorphism affecting size may also occur in ostracods, and the work of Reyment and Van Valen (1969), for example, illustrates the complications that may arise when both dimorphism and successive instars have to be considered.

More generally, of course, post-mortal modifications must be invoked, because most benthonic animals are either sessile or capable of only limited migration. Six possibilities are recognized.

Collection failure. Without doubt many museum collections are deficient in juvenile specimens because they were ignored by the collectors. Too many paleontologists have actively sought in vain for juveniles in their field and laboratory work, however, for this factor to account for more than special instances.

Current transport. Since water currents can be observed to transport shells it is not surprising that the notion of normal size-frequency distributions being primarily the result of selective removal of small shells has proved a popular one (e.g., Boucot, 1953; Fagerstrom, 1964). This may well happen in some instances, though reliable documentation is lacking (Hallam, 1967). One might reasonably ask where the small shells eventually ended up, because they ought to be found somewhere. In actuality many fine-grained deposits, in which small shells might have been expected to settle, are just as deficient in them as are more coarse-grained beds (e.g., Broadhurst, 1964). Where, moreover, the process of selective shell transport has been observed in the littoral regime unexpected results have sometimes emerged. Thus Lever *et al.* (1964) found that large valves of *Donax vittatus* are actually transported along the Dutch coast more readily than small valves, and Wilson (1967) observed that *Cerastoderma edule* valves of intermediate size were preferentially removed by currents, leaving behind a bimodal size distribution of smaller and larger shells buried by scouring.

The possibility that size-selective current transport has taken place can be put to test in another way. I made a study of two obvious death assemblages consisting of disarticulated and transported valves of many molluscan species, one from the strandline of the Firth of Forth, Scotland, and the other from the Pleistocene of Newport Bay, California (Hallam, 1967). The various species ranged widely in size, and had a variety of size-frequency distributions. Some larger-sized species were deficient in juveniles, but minute shells of smaller-sized species were numerous. If small size grades had been selectively removed by current transport, the latter should also have been affected. Other work on modern coastal regions has tended to discount the importance of selective current transport in most of the environmental regimes studied (Craig, 1967; Cadée, 1968).

Mechanical destruction. The tumbling barrel experiments performed by Chave (1964) showed clearly the wide variability in resistance to destruction of the skeletons of different phyla. Similar experiments on shells of the same species of bivalve and gastropod demonstrated a striking size-selective effect in the former, the smaller shells being rapidly destroyed (Hallam, 1967). Mechanical destruction could be an important factor therefore in environments of high energy, where shells are subjected to frequent impact by rolling and bouncing. Driscoll (1970) has estimated for several bivalve species that shell destruction is 150 to 1000 times more rapid in surf zones than in low-energy sublittoral environments. The same worker, however (1967), reckons that surf action modifies bivalve shells much more slowly than laboratory abrasion in tumbling barrels, a view with which I strongly concur.

Solution. It is quite easy to envisage selective solution of thin, minute shells of certain small-sized species during diagenesis; whether it actually occurs on a large scale remains to be demonstrated. This factor can certainly be discounted for many species of larger-sized and comparatively robust calcitic shells, and for many fine-grained rocks. It is probably in general of minor importance.

Predation. Though a proportion (usually a minority) of, for instance, bivalve shells contain holes drilled by carnivorous gastropods, no pronounced size-selective effect of the sort required has been clearly demonstrated.

Growth and mortality rates. Population dynamic analysis of fossil assemblages and their Recent equivalents leaves little doubt that the major factors molding size-frequency distributions are growth and mortality rates. Year-to-year fluctuations in recruitment should average out so that fossil assemblages, though giving only a blurred picture, should normally reflect mean growth and mortality rates. A moderate positive skewness rather than a normal distribution seems to be the more usual pattern, and negative skewness is highly exceptional, in fossils as for Recent organisms, the only case I know being one recorded by McKerrow *et al.* (1969) for a Jurassic brachiopod species. A paradox remains. Judging by what is known of post-larval mortality rates, many more distributions than are found should show strong positive skewness. Perhaps the explanation lies partly in the fact that ecologists interested in population dynamics of the marine benthos have largely confined their attention to the relatively disturbed environments of the littoral and shallow sublittoral zone, whereas most assemblages preserved in the fossil record probably lived in slightly deeper and quieter zones where the hazards of growing up were somewhat less.

Long-term fluctuations and regional variations. No systematic study has yet been attempted of long-term population fluctuations in the fossil record, though I and many others have often been struck by the way in which one observes, ascending a stratigraphic sequence, first one species and then another springing into prominence by increasing greatly in abundance, and then diminishing just as markedly after a brief interval. Such fluctuations might be repeated many times before a given species disappears for the last time, and do not appear to show any notable correlations either with other organisms or with the sedimentary facies. They almost certainly correspond with the long-term fluctuations that are known to occur at the present day. Careful quantitative work on favorable sequences might well allow a distinction to be made between equilibrium and opportunistic species. Levinton (1970) suggests that the latter might be recognizable by their tendency to occur as clusters or in thin beds where they may dominate a variety of otherwise distinct faunal assemblages to the extent of 85% or more.

Sanders (1968) and others have discussed how diversity of benthonic invertebrates often varies inversely with faunal density (i.e., individual abundance). Thus biologically accommodated communities in stable, predictable environments are characterised by high diversity and low density, while the reverse holds for the communities in unpredictable environments. As we have seen, this difference can be interpreted in terms of *K* and *r* selection, respectively. We can perceive from this a ready explanation for a phenomenon observed repeatedly in the stratigraphic record, viz., the occurrence of low-diversity faunas in marginal marine or brackish water environments where a small number of invertebrate species occur in huge quantities, often crowding whole bedding planes over many square meters. The possibility of stratigraphic condensation can often be ruled out in such cases, where the beds in question equal or exceed in thickness lateral correlatives in more normal marine facies.

We may suspect from such cases that many of these species were behaving "opportunistically," whether they were eurytopic marine or genetically adapted to unpredictably or strongly fluctuating environments. Full allowance must therefore be made by the paleoecologist for this phenomenon in his interpretation. From personal observations on the European Jurassic I would suspect that *r* selection became important in a variety of stressful environments such as low or high salinity, low temperature, strongly fluctuating salinity or temperature, low oxygen content and perhaps high turbidity.

It is certainly arguable whether or not colonizing episodes of the sort discussed by MacArthur and Wilson (1967) are too short-lived to leave any perceptible trace in the fossil record, save perhaps in highly exceptional circumstances. I can think of at least one type of situation where the colonizing may be sufficiently gradual to leave a good fossil record, namely episodes of slow transgression of epicontinental seas. In such cases non-marine deposits or basement may be overlain by a series of beds containing faunas of high density and low diversity, which pass up progressively into beds with faunas of lower density and higher diversity. The transition from the Jurassic (Bathonian) Great Estuarine Series of the Scottish Hebrides into the Callovian-Oxfordian Staffin Shales is one good example. Another is the Lower Liassic sequence in Portugal (Hallam, 1971). (Other transgressive phases are admittedly only represented by condensed shell beds.)

Future Work

As in many other fields of paleoecology we are currently handicapped by insufficiency of knowledge so that data are often too sketchy to allow more than the most tentative generalizations. Fuller analysis of many more living species is required, especially of mortality rates in quiet subtidal regimes and the effects of current transport and mechanical destruction of shells.

As regards fossils, attention should be concentrated on those rocks from which all specimens can be extracted, such as shales, clays and unconsolidated sands, to avoid sampling bias in favor of larger specimens. To establish norms a vast amount of size-frequency determinations is required. Where fossiliferous sands and clays alternate, possible differences in size distribution should be investigated. Many of the so-called dwarfed faunas of black shales might consist of young forms, since high juvenile mortality could be expected under conditions of oxygen deficiency (Hallam, 1965). This could be tested by thorough size-frequency analysis, which might reveal a small proportion of much larger adults, in conjunction with discernment of criteria of maturity.

How often does sexual dimorphism in ammonites show up as bimodal size distributions, and if not, why not? Are the smaller ammonites of a given species mature microconchs, juveniles or incomplete nuclei? Are moulting groups such as ostracods and trilobites usually characterized by distinctive size-frequency distributions which reveal the instars? Can temperature gradients be determined from growth and size data? Which fossil groups, if any, were more liable to long-term population fluctuations than others, and what might this signify? Can K and r selection be reasonably inferred in the fossil record, and if so what light might this throw on biological relationships or the physical environment?

These are just some of the questions that the population dynamics approach can illuminate, if not resolve. Taken in conjunction with other data, the analysis of size distributions and abundance offers much promise for the paleoecologist.

5

PUNCTUATED EQUILIBRIA: AN ALTERNATIVE TO PHYLETIC GRADUALISM

Niles Eldredge • *Stephen Jay Gould*

Editorial introduction. Moving from populations to species, we recall that the process of speciation as seen through the hyperopic eyes of the paleontologist is an old and venerable theme. But the significance of "gaps" in the fossil record has been a recurrent "difficulty," used on the one hand to show that spontaneous generation is a "fact," and on the other hand to illustrate the "incompleteness" of the fossil record. Some have expressed a third interpretation, which views such gaps as the logical and expected result of the allopatric model of speciation.

Bernard's Eléments de Paléontologie (1895) discusses the existence of gaps in the fossil record as follows, p. 25, English edition:

> Still it remains an indisputable fact that in the most thoroughly explored regions, those where the fauna is best known, as, for instance, the Tertiary of the Paris basin, the species of one bed often differ widely from those of the preceding, even where no stratigraphic gap appears between them. This is easily explained. The production of new forms usually takes place within narrowly limited regions. It may happen in reality that one form evolves in the same manner in localities widely separated from each other, and farther on we shall see examples of this; but this is not generally the case, the area of the appearance of species is

usually very circumscribed. This fact has been established in the case of certain butterflies and plants. The diversity having once occurred, the new types spread often to great distances, and may be found near the present form without crossing with it or presenting any trace of transition.

The same phenomenon must have taken place in former epochs. It is then only by the merest chance that geologists are able to locate the origin of the species they have under consideration; if, furthermore, the phenomena of erosion or metamorphism have destroyed or changed the locality in question, direct observation will not furnish any means of supplying the missing links of the chain.

Although this has been pointed out nicely by Bernard—and moreover, any number of paleontologists will tell you that this is what they teach—comprehension and application are two different things. And indeed, the fossil record has been interpreted by many to show just the opposite. J. B. S. Haldane's classical *The Cause of Evolution*, published in 1932, contains the following passage (p. 213):

> But [Sewall] Wright's theory [that evolution is most rapid in populations large enough to be reasonably variable, but small enough to permit large changes in gene frequencies due to random drift] certainly supports the view taken in this book that the evolution in large random-mating populations, which is recorded by paleontology, is not representative of evolution in general, and perhaps gives a false impression of the events occurring in less numerous species.

Thus an extremely eminent student of the evolutionary process considered that the known fossil record supported the view of evolution proceeding as a stately unfolding of changes in large populations.

The interpretation supported by Eldredge and Gould is that allopatric speciation in small, peripheral populations *automatically* results in "gaps" in the fossil record. Throughout their essay, however, runs a larger and more important lesson: *a priori* theorems often determine the results of "empirical" studies, before the first shred of evidence is collected. This idea, that theory dictates what one sees, cannot be stated too strongly.

Statement

In this paper we shall argue:

(1) The expectations of theory color perception to such a degree that new notions seldom arise from facts collected under the influence of old pictures of the world. New pictures must cast their influence before facts can be seen in different perspective.

(2) Paleontology's view of speciation has been dominated by the picture of "phyletic gradualism." It holds that new species arise from the slow and steady transformation of entire populations. Under its influence, we seek unbroken fossil series linking two forms by insensible gradation as the only complete mirror of Darwinian processes; we ascribe all breaks to imperfections in the record.

(3) The theory of allopatric (or geographic) speciation suggests a different interpretation of paleontological data. If new species arise very rapidly in small, peripherally isolated local populations, then the great expectation of insensibly graded fossil sequences is a chimera. A new species does not evolve in the area of its ancestors; it does not arise from the slow transformation of all its forbears. Many breaks in the fossil record are real.

(4) The history of life is more adequately represented by a picture of "punctuated equilibria" than by the notion of phyletic gradualism. The history of evolution is not one of stately unfolding, but a story of homeostatic equilibria, disturbed only "rarely" (i.e., rather often in the fullness of time) by rapid and episodic events of speciation.

The Cloven Hoofprint of Theory

> Innocent, unbiased observation is a myth.
>
> P. B. Medawar (1969, p. 28)

Isaac Newton possessed no special flair for the turning of phrases. Yet two of his epigrams have been widely cited as guides for the humble and proper scientist—his remark in a letter of 1675 written to Hooke: "If I have seen farther, it is by standing on the shoulders of giants," and his confusing comment of the *Principia* (1726 edition, p. 530): "hypotheses non fingo"—[I frame no hypotheses]. The first is not his own; it has a pedigree extending back at least to Bernard of Chartres in 1126 (Merton, 1965). The second is his indeed, but modern philosophers have offered as many interpretations for it as the higher critics heaped upon Genesis 1 in their heyday (see Mandelbaum, 1964, p. 72 for a bibliography).

Although most scholars would now hold, with Hanson (1969, 1970, see also Koyré, 1968), that Newton meant only to eschew idle speculation and untestable opinion, his phrase has traditionally been interpreted in another light—as the credo of an inductivist philosophy that views "objective" fact as the primary input to science and theory as the generalization of this unsullied information. For example, Ernst Mach, the great physicist-philosopher, wrote (1893, p. 193): "Newton's reiterated and emphatic protestations that he is not concerned with hypotheses as to the causes of phenomena, but has simply to do with the investigation and transformed statement of *actual facts* . . . stamps him as a philosopher of the *highest* rank."

Today, most philosophers and psychologists would brand the inductivist credo as naive and untenable on two counts:

(1) We do not encounter facts as *data* (literally "given") discovered objectively. All observation is colored by theory and expectation. (See Vernon, 1966, on the relation between expectation and perception. For a radical view, read Feyerabend's (1970) claim that theories act as "party lines" to force observation in preset channels, unrecognized by adherents who think they perceive an objective truth.)

(2) Theory does not develop as a simple and logical extension of observation; it does not arise merely from the patient accumulation of facts. Rather, we observe in order to test hypotheses and examine their consequences. Thus, Hanson (1970, pp. 22–23) writes: "Much recent philosophy of science has been dedicated to disclosing that a 'given' or a 'pure' observation language is a myth-eaten fabric of philosophical fiction. . . . In any observation statement the cloven hoofprint of theory can readily be detected."

Yet, inductivist notions continue to control the methodology and ethic of practicing scientists raised in the tradition of British empiricism. In unguarded moments, great naturalists have correctly attributed their success to skill in hypothesizing and power in imagination; yet, in the delusion of conscious reflection, they have usually ascribed their accomplishments to patient induction. Thus, Darwin, in a statement that should be a motto for all of us (letter to Fawcett, September 18, 1861, quoted in Medawar, 1969), wrote:

> About thirty years ago there was much talk that geologists ought only to observe and not theorize; and I well remember someone saying that at this rate a man might as well go into a gravel-pit and count the pebbles and describe the colours. How odd it is that anyone should not see that all observation must be for or against some view if it is to be of any service.

Yet, in traditional obeisance to inductivist tenets, he wrote in his autobiography that he had "worked on true Baconian principles, and without any theory collected facts on a wholesale scale" (see discussion of this point in Ghiselin, 1969a; Medawar, 1969; and de Beer, 1970).

Almost all of us adhere, consciously or unconsciously, to the inductivist methodology. We do not recognize that all our perceptions and descriptions are made in the light of theory. Leopold (1969, p. 12), for example, claimed that he could describe and analyze the aesthetics of rivers "without introduction of any personal preference or bias." He began by generating "uniqueness" values, but abandoned that approach when the sluggish, polluted, murky Little Salmon River scored highest among his samples. He then selected a very small subset of his measures for a simplified type of multivariate scaling. As he must have known before he started, Hells Canyon of the Snake River now ranked best. It cannot be accidental that the article was written by an opponent to applications then before the Federal Power Commission for the damming of Hells Canyon. (It is no less fortuitous that so many philosophers, Hegel and Spencer in particular, generated ideal states by pure reason that mirrored their own so well.)

In paleontology, even the most "objective" undertaking, the "pure" description of fossils, is all the more affected by theory because that theory is unacknowledged. We describe part by part and are led, subtly but surely, to the view that complexity is irreducible. Such description stands against a developing science of form (Gould, 1970a, 1971a) because it both gathers different facts (static states rather than dynamic correlations) and presents contrary comparisons (compendia of differences rather than reductions of complexity to fewer generating factors). D'Arcy Thompson, with his usual insight, wrote of the "pure" taxonomist (1942, p. 1036), "when comparing one organism with another, he describes the differences between them point by point and 'character' by 'character.' If he is from time to time constrained to admit the existence of 'correlation' between characters . . . yet all the while he recognizes this fact of correlation somewhat vaguely, as a phenomenon due to causes which, except in rare instances, he can hardly hope to trace; and he falls readily into the habit of thinking and talking of evolution as though it had proceeded on the lines of his own description, point by point and character by character."

The inductivist view forces us into a vicious circle. A theory often compels us to see the world in its light and support. Yet, we think we see objectively and therefore interpret each new datum as an independent confirmation of our theory. Although our theory may be wrong, we cannot confute it. To extract ourselves from this dilemma, we must bring in a more adequate theory; it will not arise from facts collected in the old way. Paleontology supported creationism in continuing comfort, yet the imposition of Darwinism forced a new, and surely more adequate, interpretation upon old facts. Science progresses more by the introduction of new world-views or "pictures"* than by the steady accumulation of information.

This issue is central to the study of speciation in paleontology. We believe that an inadequate picture has been guiding our thoughts on speciation for 100 years. We hold that its influence has been all the more tenacious because paleontologists, in claiming that they see objectively, have not recognized its guiding sway. We contend that a notion developed elsewhere, the theory of allopatric speciation, supplies a more satisfactory picture for the ordering of paleontological data.

---

* We have no desire to enter the tedious debate over what is, or is not, a "model," "theory," or "paradigm" (Kuhnian, not Rudwickian). In using the neutral word "picture," we trust that readers will understand our concern with alternate ways of seeing the world that render the same facts in *different* ways.

Phyletic Gradualism: Our Old and Present Picture

Je mehr sich das palaeontologische Material vergrössert, desto zahlreicher und vollständiger werden die Formenreihen.

Zittel, 1895, p. 11

Charles Darwin viewed the fossil record more as an embarrassment than as an aid to his theory. Why, he asked (1859, p. 310), do we not find the "infinitely numerous transitional links" that would illustrate the slow and steady operation of natural selection? "Why then is not every geological formation and every stratum full of such intermediate links? Geology assuredly does not reveal any such finely graduated organic chain; and this, perhaps, is the gravest objection which can be urged against my theory" (1859, p. 280). Darwin resolved this dilemma by invoking the great inadequacy of surviving evidence (1859, p. 342): "The geological record is extremely imperfect and this fact will to a large extent explain why we do not find interminable varieties, connecting together all the extinct and existing forms of life by the finest graduated steps. He who rejects these views on the nature of the geological record, will rightly reject my whole theory."

Thus, Darwin set a task for the new science of evolutionary paleontology: to demonstrate evolution, search the fossil record and extract the rare exemplars of Darwinian processes—insensibly graded fossil series, spared somehow from the ravages of decomposition, non-deposition, metamorphism, and tectonism. Neither the simple testimony of change nor the more hopeful discovery of "progress" would do, for anti-evolutionists of the catastrophist schools had claimed these phenomena as consequences of their own theories. The rebuttal of these doctrines and the test for (Darwinian) evolution could only be an *insensibly graded fossil sequence*—this discovery of all transitional forms linking an ancestor with its presumed descendant (*figure 5–1*). The task that Darwin set has guided our studies of evolution to this day.*

In titling his book *On the Origin of Species by Means of Natural Selection*, Darwin both identified this event as the keystone of evolution and stated his belief in its manner of occurrence. New species can arise in only two ways: by the transformation of an entire population from one state to another (phyletic evolution) or by the splitting of a lineage (speciation). The second process must occur: otherwise there could be no increase in numbers of taxa and life would cease as lineages became extinct. Yet, as Mayr (1959) noted, Darwin muddled this distinction and cast most of his discussion in terms of phyletic

---

* Beliefs in "saltative" evolution, buttressed by de Vries' "mutation theory," collapsed when population geneticists of the 1930's welded modern genetics and Darwinism into our "synthetic theory" of evolution. The synthetic theory is completely Darwinian in its identification of natural selection as the efficient cause of evolution.

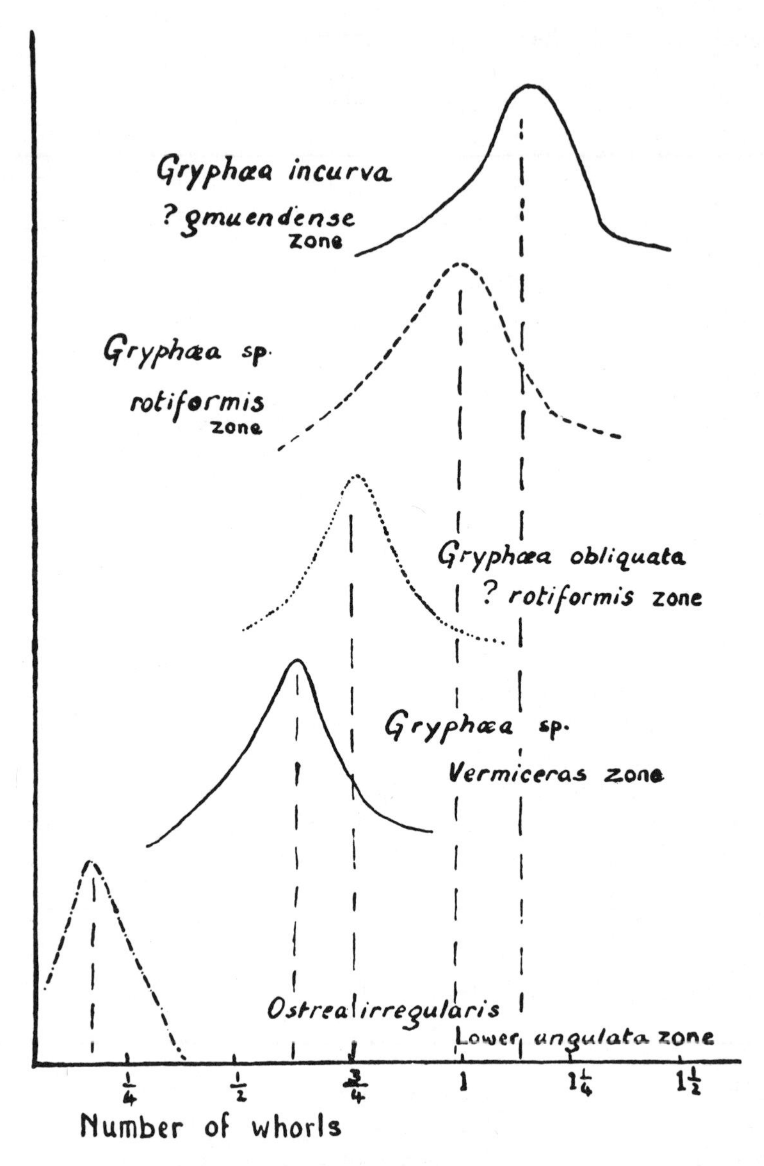

Figure 5–1: The classic case of postulated phyletic gradualism in paleontology. Slow, progressive, and gradual increase in whorl number in the basal Liassic oyster *Gryphaea.* From Trueman, 1922; figure 5.

evolution. His insistence on insensibly graded sequences among fossils reflects this emphasis, for if species arise by the gradual transformation of entire populations, an even sequence of intermediates should indeed be found. When Darwin did discuss speciation (the splitting of lineages), he continued to look through the glasses of transformation: he saw splitting largely as a sympatric process, proceeding slowly and gradually, and producing progressive divergence between forms. To Darwin, therefore, speciation entailed the same expectation as phyletic evolution: a long and insensibly graded chain of intermediate forms. Our present texts have not abandoned this view (*figure 5–2*), although modern biology has.

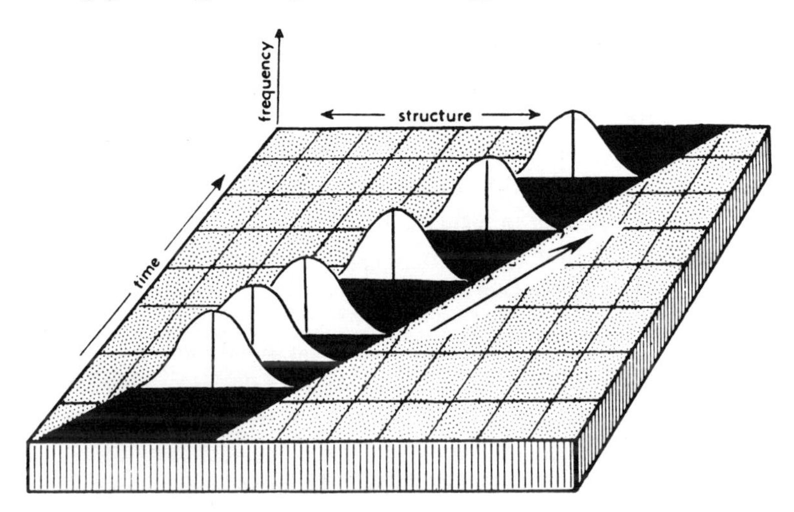

Figure 5–2: A standard textbook view of evolution *via* phyletic gradualism. From Moore, Lalicker, and Fischer, 1952; figure 1–14.

In this Darwinian perspective, paleontology formulated its picture for the origin of new taxa. This picture, though rarely articulated, is familiar to all of us. We refer to it here as "phyletic gradualism" and identify the following as its tenets:

(1) New species arise by the transformation of an ancestral population into its modified descendants.

(2) The transformation is even and slow.

(3) The transformation involves large numbers, usually the entire ancestral population.

(4) The transformation occurs over all or a large part of the ancestral species' geographic range.

These statements imply several consequences, two of which seem especially important to paleontologists:

(1) Ideally, the fossil record for the origin of a new species should consist of a long sequence of continuous, insensibly graded intermediate forms linking ancestor and descendant.

(2) Morphological breaks in a postulated phyletic sequence are due to imperfections in the geological record.

Under the influence of phyletic gradualism, the rarity of transitional series remains as our persistent bugbear. From the reputable claims of a Cuvier or an Agassiz to the jibes of modern cranks and fundamentalists, it has stood as the bulwark of anti-evolutionist arguments: "For evolution to be true, there had to be thousands, millions of transitional forms making an unbroken chain" (Anon., 1967—from a Jehovah's Witnesses pamphlet).

We have all heard the traditional response so often that it has become imprinted as a catechism that brooks no analysis: the fossil record is extremely imperfect. To cite but one example: "The connection of arbitrarily selected 'species' in a time sequence, in fact their complete continuity with one another, is to be expected in all evolutionary lineages. But, *fortunately*, because of the imperfect preservation of fossil faunas and floras, we shall meet relatively few examples of this, no matter how long paleontology continues" (Eaton, 1970, p. 23—our italics; we are amused by the absurdity of a claim that we should rejoice in a lack of data because of the taxonomic convenience thus provided).

This traditional approach to morphological breaks merely underscores what Feyerabend meant (see above) in comparing theories to party lines, for it renders the picture of phyletic gradualism virtually unfalsifiable. The picture prescribes an interpretation and the interpretation, viewed improperly as an "objective" rendering of data, buttresses the picture. We have encountered no dearth of examples, and cite the following nearly at random. Neef (1970) encountered "apparent saltation in the *Pelicaria* lineage" (p. 464), a group of Plio-Pleistocene snails. Although he cites no lithologic or geographic data favoring either interpretation, the picture of phyletic gradualism prescribes a preference: "It is likely that the discontinuity . . . is due to a period of non-deposition. . . . The possibility that the apparent saltations in the *Pelicaria* lineage are due to the migration of advanced forms from small nearby semi-isolated populations and that deposition of the Marima Sandstone was continuous cannot be entirely excluded" (1970, p. 454).

Moreover, the picture's influence has many subtle extensions. For instance:

(1) It colors our language. We are compelled to talk of "morphological breaks" in order to be understood. But the term is not a neutral descriptor; it presupposes the truth of phyletic gradualism, for a "break" is an interruption of something continuous. (Under a deVriesian picture, for example, "breaks" are "saltations"; they are real and expressive of evolutionary processes.)

(2) It prescribes the cases that are worthy of study. If breaks are artificial, the sequences in which they abound become, *ipso facto*, poor objects for evolutionary investigation. But surely there is something insidious here: if breaks are real and stand against the picture of phyletic gradualism, then the picture itself excludes an investigation of the very cases that could place it in jeopardy.

If we doubt phyletic gradualism, we should not seek to "disprove" it "in the rocks." We should bring a new picture from elsewhere and see if it provides a more adequate interpretation of fossil evidence. In the next section, we express our doubts, display a different picture, and attempt this interpretation.

But before leaving the picture of phyletic gradualism, we wish to illustrate its pervasive influence in yet another way. Kuhn (1962) has stressed the impact of textbooks in molding the thought of new professionals. The "normal science" that they inculcate is "a strenuous and devoted attempt to force nature into the conceptual boxes supplied by professional education" (1962, p. 5).

Before the "modern synthesis" of the 1930's and 40's, English-speaking invertebrate paleontologists were raised upon two texts—Eastman's translation of Zittel (1900) and that venerable *Gray's Anatomy* of British works, Woods' *Palaeontology* (editions from 1893 to 1946, last edition reprinted five times before 1958 and still very much in use). Both present an orthodox version of phyletic gradualism. In a classic statement, ending with the sentence that serves as masthead to this section, Zittel wrote (Eastman translation, 1900, p. 10):

> Weighty evidence for the progressive evolution of organisms is afforded by fossil transitional series, of which a considerable number are known to us, notwithstanding the imperfection of the palaeontological record. By transitional series are meant a greater or lesser number of similar forms occurring through several successive horizons, and constituting a practically unbroken morphic chain . . . With increasing abundance of palaeontological material, the more numerous and more complete are the series of intermediate forms which are brought to light.

The last edition of Woods (1946) devotes three pages to evolution; all but two paragraphs (one on ontogeny, the other on orthogenesis) to an exposition of phyletic gradualism (one page on the imperfection of the record, another on some rare examples of graded sequences).

Our current textbooks have changed the argument not at all. Moore, Lalicker and Fischer (1952, p. 30), in listing the fossil record among "evidences of evolution," have only this to say about it: "Although lack of knowledge is immeasurably greater than knowledge, many lineages among fossils of various groups have been firmly established. These demonstrate the transformation of one species or genus into another and thus constitute documentary evidence of gradual evolution." And Easton (1960, p. 34), citing the apotheosis of our achievements, writes: "An evolutionary series represents the peak of scientific accomplishment in organizing fossil invertebrates. It purports to show an orderly progression in morphologic changes among related creatures during successive intervals of time."

That these older texts hold so strongly to phyletic gradualism should surprise no one; harder to understand is the fact that virtually all modern texts repeat the same arguments even though their warrant had disappeared, as we shall now show, with the advent of the allopatric theory of speciation.

The Biospecies and Punctuated Equilibria: A Different Picture of Speciation

> Habits of thought in the tradition of a science are not readily changed, it is not easy to deviate from the customary channels of accumulated experience in conventionalized subjects.
>
> G. L. Jepsen, 1949, p. v

An irony. The formulation of the biological species concept was a major triumph of the synthetic theory (Mayr, 1963, abridged and revised 1970, remains the indispensable source on its meaning and implications). Since paleontology has always taken its conceptual lead from biology (with practical guidance from geology), it was inevitable that paleontologists should try to discover the meaning of the biospecies for their own science.

Here we meet an ironic situation: the taxonomic perspective—one of our persistent albatrosses—dictated an approach to the biospecies. Instead of extracting its insights about evolutionary processes, we sought only its prescriptions for classification. We learned that species are populations, that they are recognized in fossils by ranges of variability not by correspondence to idealized types. The "new systematics" ushered in the revolution in species-level classification that Darwin's theory had implied but not effected. In paleontology, its main accomplishment has been a vast condensation and elimination of spurious taxa established on typological criteria.

But the new systematics also rekindled a theoretical debate unsurpassed in the annals of paleontology for its ponderous emptiness: What is the nature of a paleontological species? In this reincarnation: can taxa designated as biospecies be recognized from fossils? Biologists insisted that the biospecies is a "real" unit of nature, a population of interacting individuals, reproductively isolated from all other groups. Yet its reality seemed to hinge upon what Mayr calls its "non-dimensional" aspect: species are distinct at any moment in time, but the boundaries between forms must blur in temporal extension—a continuous lineage cannot be broken into objective segments. Attempts to reconcile or divorce the non-dimensional biospecies and the temporal "paleospecies" creep on apace (Imbrie, 1957; Weller, 1961; McAlester, 1962; Shaw, 1969; and an entire symposium edited by Sylvester-Bradley, 1956); if obfuscation is any sign of futility, we offer the following as a plea for the termination of this discussion: "Such a plexiform lineage . . .

constitutes a chronospecies (or paleospecies), and it is composed of many successional polytypic morphospecies ('holomorphospecies'), each of which is in theory the paleontological equivalent of a neontological biospecies" (Thomas, 1956, p. 24).

The discussion is futile for a very simple reason: the issue is insoluble; it is not a question of fact (phylogeny proceeds as it does no matter how we name its steps), but a debate about ways of ordering information. When Whitehead said that all philosophy was a footnote to Plato, he meant not only that Plato had identified all the major problems, but also that the problems were still debated because they could not be solved. The point is this: the hierarchical system of Linnaeus was established for his world: a world of discrete entities. It works for the living biota because most species are discrete at any moment in time. It has no objective application to evolving continua, only an arbitrary one based on subjective criteria for division. Linnaeus would not have set up the same system for our world. As Vladimir Nabokov writes in *Ada* (1969, p. 406): "Man . . . will never die, because there may never be a taxonomical point in his evolutionary progress that could be determined as the last stage of man in the cline turning him into Neohomo, or some horrible throbbing slime."

Then does the biospecies offer us nothing but semantic trouble? On one level, the answer is no because it can be applied with great effectiveness to past time-planes. But on another level, and this involves our irony, we must avoid the narrow approach that embraces a biological concept only when it can be transplanted bodily into our temporal taxonomy. The biospecies abounds with implications for the operation of evolutionary processes. Instead of attempting vainly to name successional taxa objectively in its light (McAlester, 1962), we should be applying its concepts. In the following section, we argue that one of these concepts—the theory of allopatric speciation—might reorient our picture for the origin of taxa.

Implications of allopatric speciation for the fossil record. We wish to consider an alternate picture to phyletic gradualism; it is based on a theory of speciation that arises from the behavior, ecology, and distribution of modern biospecies. First, we must emphasize that mechanisms of speciation can be studied directly only with experimental and field techniques applied to living organisms. No theory of evolutionary mechanisms can be generated directly from paleontological data. Instead, theories developed by students of the modern biota generate predictions about the course of evolution in time. With these predictions, the paleontologist can approach the fossil record and ask the following question: Are observed patterns of geographic and stratigraphic distribution, and apparent rates and directions of morphological change, consistent with the consequences of a particular theory of speciation? We can apply and test, but we cannot generate new mechanisms. If discrepancies are found between paleontological data and the expected patterns, we may be able to identify those aspects of a general theory that need improvement. But we cannot formulate these improvements ourselves.*

During the past thirty years, the allopatric theory has grown in popularity to become, for the vast majority of biologists, *the* theory of speciation. Its only serious challenger is the sympatric theory. Here we discuss only the implications of the allopatric theory for interpreting the fossil record of sexually-reproducing metazoans. We do this simply because it is the allopatric, rather than the sympatric, theory that is preferred by biologists. We shall therefore contrast the allopatric theory with the picture of phyletic gradualism developed in the last section.

Most paleontologists, of course, are aware of this theory, but the influence of phyletic gradualism remains so strong that discussions of geographic speciation are almost always cast in its light: geographic speciation is seen as the slow and steady transformation of two separated lineages—i.e., as *two* cases of phyletic gradualism (*figure 5–3*). Raup and Stanley (1971, p. 98), for example, write:

> Let us consider populations of a species living at a given time but not in geographic contact with each other. . . . Two or more segments of the species thus evolve and undergo *phyletic* speciation independently. . . . The distinction between phyletic and geographic speciation is to some extent artificial in that both processes depend on natural selection. The critical difference is that phyletic speciation is accomplished in the absence of geographic isolation and geographic speciation requires geographic isolation (italics ours).

The central concept of allopatric speciation is that new species can arise only when a small local population becomes isolated at the margin of the geographic range of its parent species. Such local populations are termed *peripheral isolates*. A peripheral isolate develops into a new species if *isolating mechanisms* evolve that will prevent the re-initiation of gene flow if the new form re-encounters its ancestors at some future time. As a consequence of the allopatric theory, new fossil species do not originate in the place where their ancestors lived. It is extremely improbable that we shall be able to trace the gradual splitting of a lineage merely by following a certain species up through a local rock column.

Another consequence of the theory of allopatric processes follows: since selection always maintains an equilibrium between populations and their local environment, the morphological features that distinguish the descendant

---

* The rate and direction of morphological change over long periods of time is the most obvious kind of evolutionary pattern that we can test against predictions based on processes observed over short periods of time by neontologists. We try to do this in the next section.

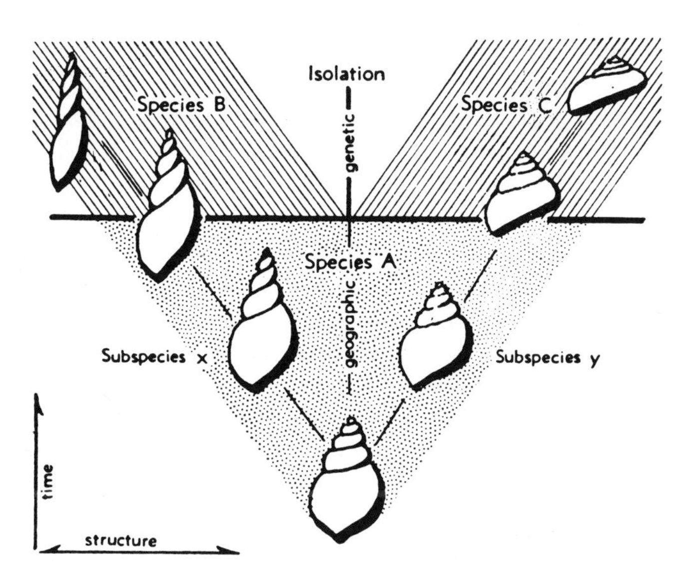

Figure 5–3: A hypothetical case of geographic speciation viewed from the perspective of phyletic gradualism—slow and gradual transformation in two lineages. From Moore, Lalicker, and Fischer, 1952; figure 1–15.

species from its ancestor are present close after, if not actually prior to, the onset of genetic isolation. These differences are often accentuated if the two species become sympatric at a later date (character displacement—Brown and Wilson, 1956). In any event, most morphological divergence of a descendant species occurs very early in its differentiation, when the population is small and still adjusting more precisely to local conditions. After it is fully established, a descendant species is as unlikely to show gradual, progressive change as is the parental species. Thus, in the fossil record, we should not expect to find gradual divergence between two species in an ancestral-descendant relationship. Most evolutionary changes in morphology occur in a short period of time relative to the total duration of species. After the descendant is established as a full species, there will be little evolutionary change except when the two species become sympatric for the first time.

These simple consequences of the allopatric theory can be combined into an expected pattern for the fossil record. Using stratigraphic, radiometric, or biostratigraphic criteria (for organisms other than those under study), we establish a regional framework of correlation. Starting with these correlations, patterns of geographic (not stratigraphic) variation among samples of fossils should appear. Tracing a fossil species through any local rock column, so long as no drastic changes occur in the physical environment, should produce *no* pattern of constant change, but one of oscillation in mean values. Closely related (perhaps descendant) species that enter the rock column should appear suddenly and show no intergradation with the "ancestral" species in morphological features that act as inter-specific differentia. There should be no gradual divergence between the two species when both persist for some time to higher stratigraphic levels. Quite the contrary—it is likely that the two species will display their greatest difference when the descendant first appears. Finally, in exceptional circumstances, we may be able to identify the general area of the ancestor's geographic range in which the new species arose.

Another conclusion is that time and geography, as factors in evolution, are not so comparable as some authors have maintained (Sylvester-Bradley, 1951). The allopatric theory predicts that most variation will be found among samples drawn from different geographic areas rather than from different stratigraphic levels in the local rock column. The key factor is adjustment to a heterogeneous series of micro-environments vs. a general pattern of stasis through time.

In summary, we contrast the tenets and predictions of allopatric speciation with the corresponding statements of phyletic gradualism previously given:

(1) New species arise by the splitting of lineages.

(2) New species develop rapidly.

(3) A small sub-population of the ancestral form gives rise to the new species.

(4) The new species originates in a very small part of the ancestral species' geographic extent—in an isolated area at the periphery of the range.

These four statements again entail two important consequences:

(1) In any *local* section containing the ancestral species, the fossil record for the descendant's origin should consist of a sharp morphological break between the two forms. This break marks the migration of the descendant, from the peripherally isolated area in which it developed, into its ancestral range. Morphological change in the ancestor, even if directional in time, should bear no relationship to the descendant's morphology (which arose in response to local conditions in its isolated area). Since speciation occurs rapidly in small populations occupying small areas far from the center of ancestral abundance, we will rarely discover the actual event in the fossil record.

(2) Many breaks in the fossil record are real; they express the way in which evolution occurs, not the fragments of an imperfect record. The sharp break in a local column accurately records what happened in that area through time. Acceptance of this point would release us from a self-imposed status of inferiority among the evolutionary sciences. The paleontologist's gut-reaction is to view almost any anomaly as an artifact imposed by our institutional millstone—an imperfect fossil record. But just as we now tend to view the rarity of Precambrian metazoans as a true reflection of life's history rather than a testimony to the ravages of metamorphism or the lacunae of Lipalian

intervals, so also might we reassess the smaller breaks that permeate our Phanerozoic record. We suspect that this record is much better (or at least much richer in optimal cases) than tradition dictates.

Problems of phyletic gradualism. In our alternate picture of phyletic gradualism, we are not confronted with a self-contained theory from modern biology. The postulated mechanism for gradual uni-directional change is "orthoselection," usually viewed as a constant adjustment to a uni-directional change in one or more features of the physical environment. The concept of orthoselection arose as an attempt to remove the explanation of gradual morphological change from the realm of metaphysics ("orthogenesis"). It does *not* emanate from *Drosophila* laboratories, but represents a hypothetical extrapolation of selective mechanisms observed by geneticists.

Extrapolation of gradual change under selection to a complete model for the origin of species fails to recognize that speciation is primarily an ecological and geographic process. Natural selection, in the allopatric theory, involves adaptation to local conditions and the elaboration of isolating mechanisms. Phyletic gradualism is, in itself, an insufficient picture to explain the origin of diversity in the present, or any past, biota.

Although phyletic gradualism prevails as a picture for the origin of new species in paleontology, very few "classic" examples purport to document it. A few authors (MacGillavry, 1968, Eldredge, 1971) have offered a simple and literal interpretation of this situation: *in situ*, gradual, progressive evolutionary change is a rare phenomenon. But we usually explain the paucity of cases by a nearly-ritualized invocation of the inadequacy of the fossil record. It *is* valid to point out the rarity of thick, undisturbed, highly fossiliferous rock sections in which one or more species occur continuously throughout the sequence. Nevertheless, if most species evolved according to the tenets of phyletic gradualism, then, no matter how discontinuous a species' occurrence in thick sections, there should be a shift in one or more variables from sample to sample up the section. This is, in fact, the situation in most cases of postulated gradualism: the "gradualism" is represented by dashed lines connecting known samples. This procedure provides an excellent example of the role of preconceived pictures in "objectively documented" cases. One of the early "classics" of phyletic gradualism, Carruthers' (1910) study of the Carboniferous rugose coral *Zaphrentites delanouei* (Milne-Edwards and Haime) and its reinterpretation by Sylvester-Bradley (1951), is of this kind. We do not say that the analysis is incorrect; the *Z. delanouei* stock may have evolved as claimed. We merely wish to show how the *a priori* picture of phyletic gradualism has imposed itself upon limited data.

How pervasive, then, is gradualism in these quasi-continuous sequences? A number of authors (including, *inter alia*, Kurtén, 1965, MacGillavry, 1968, and Eldredge, 1971) have claimed that most species show little or no change throughout their stratigraphic range. But though it is tempting to conclude that gradual, progressive morphological change is an illusion, we recognize that there is little hard evidence to support either view.

As a final, and admittedly extreme, example of *a priori* beliefs in phyletic gradualism, we cite the work of Brace (1967) on human evolution. This is all the more instructive since most paleoanthropologists, in reversing an older view that Brace still maintains, now claim that hominid evolution involves speciation by splitting as well as phyletic evolution by transformation (seen especially in the presumed coexistence of two australopithecine species in the African lower Pleistocene—Howell, 1967; Tobias, 1965; Pilbeam, 1968; Pilbeam and Simons, 1965). Brace (1967) has claimed that the fossil record of man includes four successive "stages" in direct ancestral-descendant relation. These are the Australopithecine (with two successive "phases"—the australopithecus and paranthropus), the Pithecanthropus, the Neanderthaloid, and, finally, the Modern Stage. In discussing the history of paleoanthropology, Brace shows that most denials of ancestral-descendant relationships among hominid fossils stem from a desire to avoid the conclusion that *Homo sapiens* evolved from some "lower," more "brutish" form. But Brace has lumped all such analyses under the catch phrase "hominid catastrophism." Hominid catastrophism, according to Brace, is the denial of ancestral-descendant relationships among fossils, with the invocation of extinction and subsequent migrations of new populations that arose by successive creation. Such views are, of course, absurd, but Brace would include *all* cladistic interpretations of the hominid record within "hominid catastrophism." To view hominid phylogeny as a gradual, progressive, unilineal process involving a series of stages, Brace claims, is the interpretation most consonant with evolutionary theory. His interpretation of phylogeny may be correct (though most experts deny it), but he is seriously wrong to claim that phyletic gradualism is the picture most consistent with modern biological thought. Quite apart from the issue of probable overlap in the ranges of his stages, it would be of great interest to determine the degree of stasis attained by them during any reasonably long period of time.

Application of allopatric concepts to paleontological examples. At this point, there is some justification for concluding that the picture of phyletic gradualism is poorly documented indeed, and that most analyses purporting to illustrate it directly from the fossil record are interpretations based on a preconceived idea. On the other hand, the alternative picture of stasis punctuated by episodic events of allopatric speciation rests on a few general statements in the literature and a wealth of informal data. The idea of *punctuated equilibria* is just as much a preconceived picture as that of phyletic gradualism. We readily admit our bias towards it and urge readers, in the ensuing discussion, to remember that our interpretations are as colored by our preconceptions as are the claims of the champions of phyletic gradualism by theirs. We merely reiterate: (1) that one must have some picture of speciation in mind, (2) that the data of paleontology cannot decide which picture is more adequate, and (3) that the picture of punctuated equilibria is more in accord with the process of speciation as understood by modern evolutionists.

We could cite any number of reported sequences that fare better under notions of allopatric processes than under the interpretation of phyletic gradualism that was originally applied. This is surely true for all or part of the three warhorses of the English literature: horses themselves, the Cretaceous echinoid *Micraster*, and the Jurassic oyster *Gryphaea*. Simpson (1951) has shown that the phylogeny of horses is a luxuriant, branching bush, not the ladder to one toe and big teeth that earlier authors envisioned (Matthew and Chubb, 1921). Nichols (1959) believes that *Micraster senonensis* was a migrant from elsewhere and that it did not arise and diverge gradually from *M. cortestudinarium* as Rowe (1899) had maintained. Hallam (1959, 1962) has argued that the transition from *Liostrea* to *Gryphaea* was abrupt and that *neither* genus shows *any* progressive change through the basal Liassic zones, contrary to Trueman's claim (1922, p. 258) that: "It is doubtful whether any better example of lineage of fossil forms could be found." Gould (1971b and in press) has confirmed Hallam's conclusions. Hallam interprets the sudden appearance of *Gryphaea* as the first entry into a local rock column of a species that had evolved rapidly elsewhere. He writes (1962, p. 574): "This interpretation is more in accord with the experience of most invertebrate paleontologists who, despite continued collecting all over the world and an ever-increasing amount of research, find 'cryptogenic' genera and species far more commonly than they detect gradual trends or lineages. The sort of evolution I tentatively propose for *Gryphaea* could in fact be quite normal among the invertebrates." We agree.

We choose, rather, to present two examples from our own work which we believe are interpreted best from the viewpoint of allopatric speciation. We prefer to emphasize our own work simply because we are most familiar with it and are naturally more inclined to defend our interpretations.

Gould (1969) has analyzed the evolution of *Poecilozonites bermudensis zonatus* Verrill, a pulmonate snail, during the last 300,000 years of the Bermudian Pleistocene. The specimens were collected from an alternating sequence of wind-blown sands and red soils. Formational names, dominant lithologies, and glacial-interglacial correlations are given in *table 5–1*.

The small area and striking differentiation of stratigraphic units in the Bermudian Pleistocene permit a high degree of geographic and temporal control. *P. bermudensis* (Pfeiffer) is plentiful in all post-Belmont formations; in addition, one subspecies, *P.b. bermudensis*, is extant and available for study in the laboratory.

Distinct patterns of color banding differentiate an eastern from a western population of *P. bermudensis zonatus*. The boundary between these two groups is sharp, and there are no unambiguous cases of introgression. *P.*

bermudensis zonatus was divided into two stocks, evolving in parallel with little gene flow between them, throughout the entire interval of Shore Hills to Southampton time. Both eastern and western *P.b. zonatus* became extinct sometime after the deposition of Southampton dunes; they were replaced by *P.b. bermudensis*, a derivative of eastern *P.b. zonatus* which had been evolving separately in the area of St. George's Island since St. George's time. Gould (1969, 1970b) has discussed the parallel oscillation of several morphological features in both stocks of *P.b. zonatus*; these are adaptive shifts in response to glacially-controlled variations in climate. Both stocks exhibit stability in other features that serve to distinguish them from their nearest relatives. There is no evidence for any gradual divergence between eastern and western *P.b. zonatus*.

Several samples of *P. bermudensis* share many features that distinguish them from *P. bermudensis zonatus*. These characters can be arranged in four categories: color, general form of the spire, thickness of the shell, and shape of the apertural lip. The ontogeny of *P.b. zonatus* illustrates the interrelation of these categories. Immature shells of *P.b. zonatus* are weakly colored, relatively wide, lack a callus, and have the lowest portion of the outer apertural lip at the umbilical border. This combination of character states is exactly repeated in the large *mature* shells of non-*zonatus* samples of *P. bermudensis*. Since every ontogenetic feature developed at or after the fifth whorl in non-*zonatus* samples is attained by whorls 3–4 in *P.b. zonatus*, Gould (1969) concludes that the non-*zonatus* samples of *P. bermudensis* are derived by paedomorphosis from *P.b. zonatus*.

These paedomorphic samples range through the entire interval of Shore Hills to Recent. The most obvious hypothesis would hold that they constitute a continuous lineage evolving separately from *P.b. zonatus*. Gould rejects this and concludes that paedomorphic offshoots arose from the *P.b. zonatus* stock at four different times; the arguments are based on details of stratigraphic and geographic distrubution, as well as on morphology.

Figure 5–4 summarizes the history of splitting in the *P.b. zonatus* lineage. The earliest paedomorph, *P.b. fasolti* Gould, occurs in the Shore Hills Formation within the geographic range of eastern *P.b. zonatus*. *P.b. fasolti* and the contemporary population of eastern *P.b. zonatus* share a unique set of morphological features including, *inter alia*, small size at any given whorl, low spire, relatively wide shell, and a wide umbilicus. These features unite the Shore Hills paedomorph and non-paedomorph, and set them apart from all post-Shore Hills *P. bermudensis*.

In the succeeding Harrington Formation, paedomorphic samples of *P. bermudensis* lived in both the eastern and western geographic regions of *P.b. zonatus*. The eastern paedomorph, *P.b. sieglindae* Gould, may have evolved from the Shore Hills paedomorph, *P.b. fasolti*. However, both *P.b. sieglindae*

Table 5–1. Stratigraphic column of Bermuda.

Formation	*Description*	*Interpretation*
Recent	Poorly developed brownish soil or crust	Interglacial
Southampton	Complex of eolianites and discontinuous unindurated zones	,,
St. George's	Red paleosol of island wide extent	Glacial
Spencer's Point	Intertidal marine, beach and dune facies	Interglacial
Pembroke	Extensive eolianites and discontinuous unindurated zones	,,
Harrington	Fairly continuous unindurated layer with shallow water marine and beach facies	,,
Devonshire	Intertidal marine and poorly developed dune facies	,,
Shore Hills	Well-developed red paleosol of island-wide extent	Glacial
Belmont	Complex shallow water marine, beach and dune facies	Interglacial
Soil (?)	A reddened surface rarely seen in the Walsingham district	Glacial?
Walsingham	Highly altered elolianites	Interglacial

and the contemporaneous population of eastern *P.b. zonatus* lack the distinctive features of all Shore Hills *P. bermudensis* and a more likely hypothesis holds that the features uniting all post-Shore Hills *P. bermudensis* were evolved only once. If this is the case, *P.b. sieglindae* is a second paedomorphic derivative of eastern *P.b. zonatus*.

P.b. sieglindae differs from its contemporary paedomorph *P.b. siegmundi* Gould in that each displays the color pattern of the local non-paedomorph. Very simply, *P.b. sieglindae* is found in eastern Bermuda and shares the banding pattern of eastern *P.b. zonatus*, while *P.b. siegmundi* is found in western Bermuda and has the same color pattern as western *P.b. zonatus*. In addition, both *P.b. sieglindae* and *P.b. siegmundi* evolved at the periphery of the known range of their putative ancestors. The independent derivation of the two Harrington paedomorphs from the two stocks of *P.b. zonatus* seems clear.

Finally, the living paedomorph, *P.b. bermudensis*, first appears in the St. George's Formation on St. George's Island. While St. George's Island is within the geographic range of eastern *P.b. zonatus*, it is far removed from the

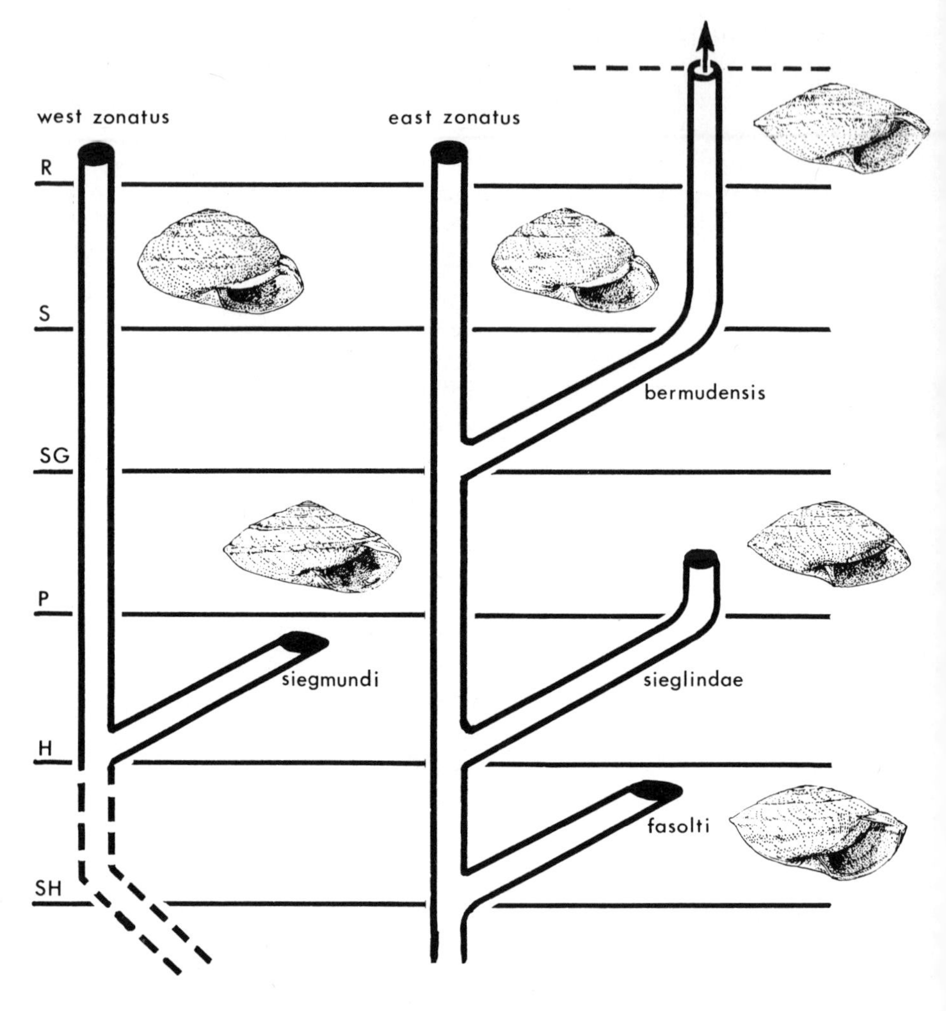

Figure 5–4: Reconstruction of the phylogenetic history of *P. bermudensis* showing iterative development of paedomorphic subspecies. SH—Shore Hills; H—Harrington; P—Pembroke; SG—St. George's; S—Southampton; R—Recent. From Gould, 1969; figure 20.

area in which *P.b. sieglindae* arose and lived. Gould concludes that *P.b. sieglindae* was a short-lived population that never enjoyed a wide geographic distribution; he estimates that the Pembroke population's range did not exceed 200 meters. Although there is little morphological evidence to support it, Gould recognizes a fourth paedomorphic subspecies, *P.b. bermudensis*, derived directly from (eastern) *P.b. zonatus*. The conclusion is based upon geographic and stratigraphic data.

Gould (1969) has advanced an adaptive explanation for the four separate origins of paedomorphic populations from *P.b. zonatus*. This explanation, based on the value of thin shells in lime-poor soils, need not be elaborated here. What is important, for our purposes, is to emphasize that the reconstruction of phylogenetic histories for the paedomorphs involves (1) attention to geographic data (the allopatric model), (2) discontinuous stratigraphic occurrence (a more literal interpretation of the fossil record), and (3) formal arguments based on morphology. It is entirely possible, from morphological data alone, to interpret the three paedomorphs of the eastern *zonatus* area as a gradational biostratigraphic series. *Figure 5–5* shows a tempting interpretation of phyletic gradualism for "lower eccentricity," an apertural

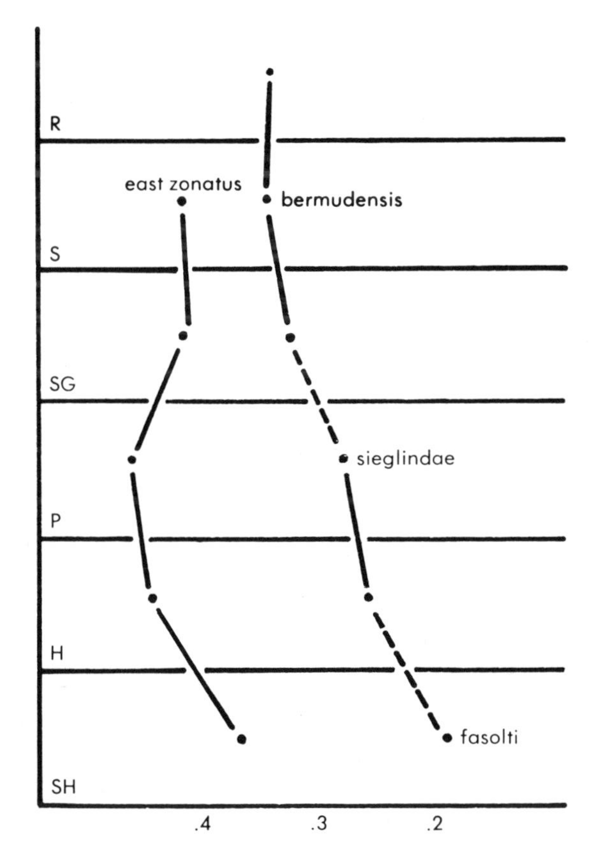

Figure 5–5: Plot of means of mean sample values of "lower eccentricity" in *P. bermudensis*. Dashed lines show the phylogeny of the three paedomorphs of eastern *zonatus* as a direct ancestral-descendant sequence, and offer a tempting instance of phyletic gradualism. Abbreviations as in *figure 5–4*.

variable. Values gradually increase through time. *Figure 5–6*, however, confounds this interpretation by showing that stratigraphic variability in "differential growth ratio" within both *P.b. sieglindae* and *P.b. bermudensis* varies in a direction *opposite* to the net stratigraphic "trend": *P.b. fasolti—P.b. sieglindae—P.b. bermudensis*: this could be read to indicate that each subspecies is unique. In fact, neither graph affords sufficient evidence to warrant either conclusion. Morphology, stratigraphy, and geography must all be evaluated.

The phylogenetic history of the trilobite *Phacops rana* (Green) from the Middle Devonian of North America (Eldredge, 1971; 1972) provides another example of the postulated operation of allopatric processes. As in *Poecilozonites bermudensis*, full genetic isolation was probably not established between "parent" and "daughter" taxa; this conclusion, based on

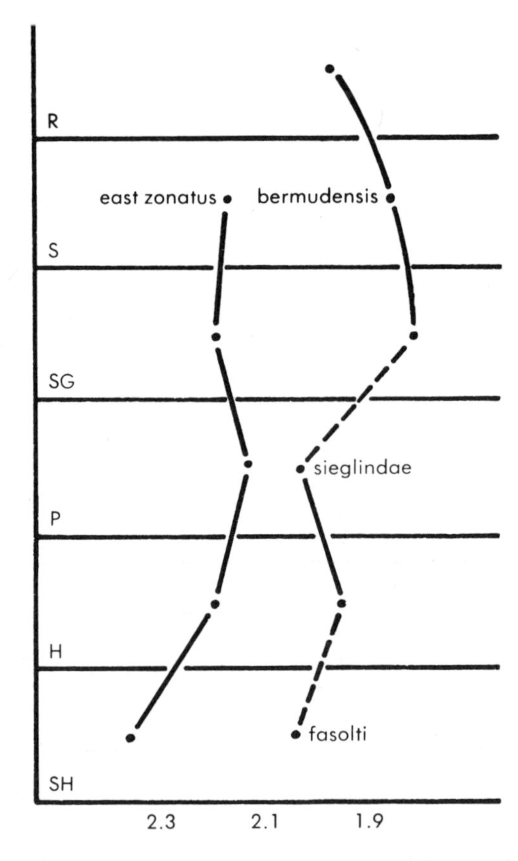

Figure 5–6: Plot of means of mean sample values for "differential growth ratio" in *P. bermudensis*. Dashed lines show the interpretation of the phylogeny of the three paedomorphs as a direct ancestral-descendant sequence. Abbreviations as in *figure 5–4*.

inferences from morphological variability, may be unwarranted. For our purposes, it does not matter whether we are dealing with four subspecies of *P. rana*, or four separate species of *Phacops*, including *P. rana* and its three closest relatives. The basic mode of evolution underlying the group's phylogenetic history as a whole is the same in either case.

Features of eye morphology exhibit the greatest amount of variation among samples of *P. rana*. Lenses are arranged on the visual surface of the eye in vertical dorso-ventral files (Clarkson, 1966). A stable number of dorso-ventral files, characteristic of the entire sample in any population, is reached early in ontogeny. The number of dorso-ventral (d.-v.) files is the most important feature of interpopulational variation in *P. rana*.

The closest known relative of *P. rana* is *P. schlotheimi* (Bronn) *s. l.*, from the Eifelian of Europe and Africa; this group has recently been revised by C. J. Burton (1969). In addition, several samples of *P. rana* have been found in the Spanish Sahara in northwestern Africa (Burton and Eldredge, in preparation). *P. schlotheimi* and the African specimens of *P. rana* are most similar to *P. rana milleri* Stewart and *P. rana crassituberculata* Stumm, the two oldest subspecies of *P. rana* in North America. All these taxa possess 18 dorso-ventral files. Eldredge (1972) concludes that 18 is the primitive number of d.-v. files for all North American *Phacops rana*.

Figure 5–7 summarizes relationships among the four subspecies of *P. rana* without regard to stratigraphic occurrence. The oldest North American *P. rana* occurs in the Lower Cazenovian Stage of Ohio and central New York State. All have 18 d.-v. files. Populations with 18 d.-v. files (*P. rana milleri* and *P. rana crassituberculata*) persist into the Upper Cazenovian Stage in the epicontinental seas west of the marginal basin in New York and the Appalachians.

Of the two samples the one that displays intra-populational variation in d.-v. file number occurs in the Lower Cazenovian of central New York. Some specimens have 18 d.-v. files, while others reduce the first d.-v. file to various degrees; a few lack it altogether. *All P. rana* from subsequent, younger horizons in New York and adjacent Appalachian states have 17 dorso-ventral files. Apparently, 17 d.-v. file *P. rana rana* arose from an 18 d.-v. file population on the northeastern periphery of the Cazenovian geographic range of *P. rana*. Seventeen d.-v. file *P. rana* persist, unchanged in most respects, through the Upper Cazenovian, Tioughniogan, and Taghanic Stages in the eastern marginal basin. Seventeen d.-v. file *P. rana rana* first appears in the shallow interior seas at the beginning of the Tioughniogan Stage, replacing the 18 d.-v. file populations that apparently became extinct during a general withdrawal of seas from the continental interior. All Tioughniogan *P. rana* possess 17 dorso-ventral files.

A second, similar event involving reduction in dorso-ventral files occurred during the Taghanic. Here again, a variable population inhabited the eastern

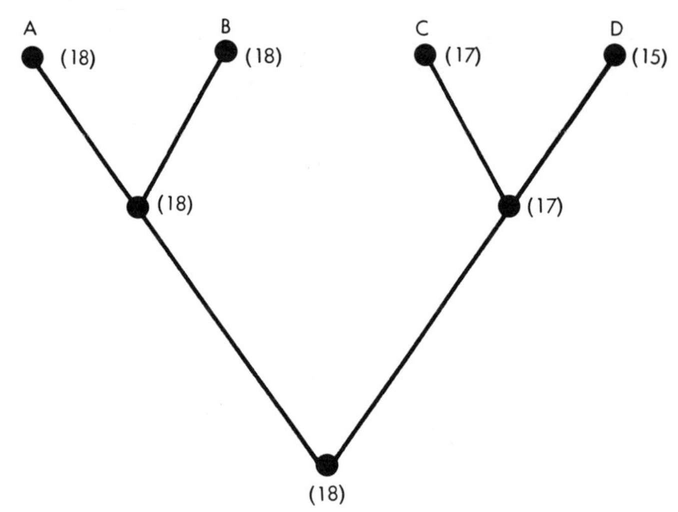

Figure 5–7: Outline of relationships of four subspecies of *Phacops rana*. A—*Phacops rana crassituberculata* Stumm; B—*Phacops rana milleri* Stewart; C—*Phacops rana rana* (Green); D—*Phacops rana norwoodensis* Stumm. Numbers in parentheses refer to number of dorso-ventral files typical of subspecies or hypothesized to characterize condition of common ancestor.

marginal basin in New York. This suggests that, once more, reduction in d.-v. files occurred allopatrically on the periphery of the known range of *P. rana rana*. The subsequent spread of stabilized, 15 d.-v. file *P. rana norwoodensis* through the Taghanic seas of the continental interior was instantaneous in terms of our biostratigraphic resolution. *Figure 5–8* summarizes this interpretation of the history of *P. rana*.

Under the tenets of phyletic gradualism, this story has a different (and incorrect) interpretation: the three successional taxa of the epeiric seas form an *in situ* sequence of gradual evolutionary modification. The sudden transitions from one form to the next are the artifact of a woefully incomplete fossil record. Most evolutionary change occurred during these missing intervals: fill in the lost pieces with an even dotted line.

If the interpreter pays attention to geographic detail, however, quite a different tale emerges, one that allows a more literal reading of the fossil record. Now the story is one of stasis: no variation in the most important feature of discrimination (number of d.-v. files—actually a complex of highly interrelated variables) through long spans of time. Two samples displaying intra-populational variation in numbers of d.-v. files identify relatively "sudden" events of reduction in files on the periphery of the species' geographic range. These two samples, moreover, have a very short stratigraphic, and very restricted geographic, distribution.

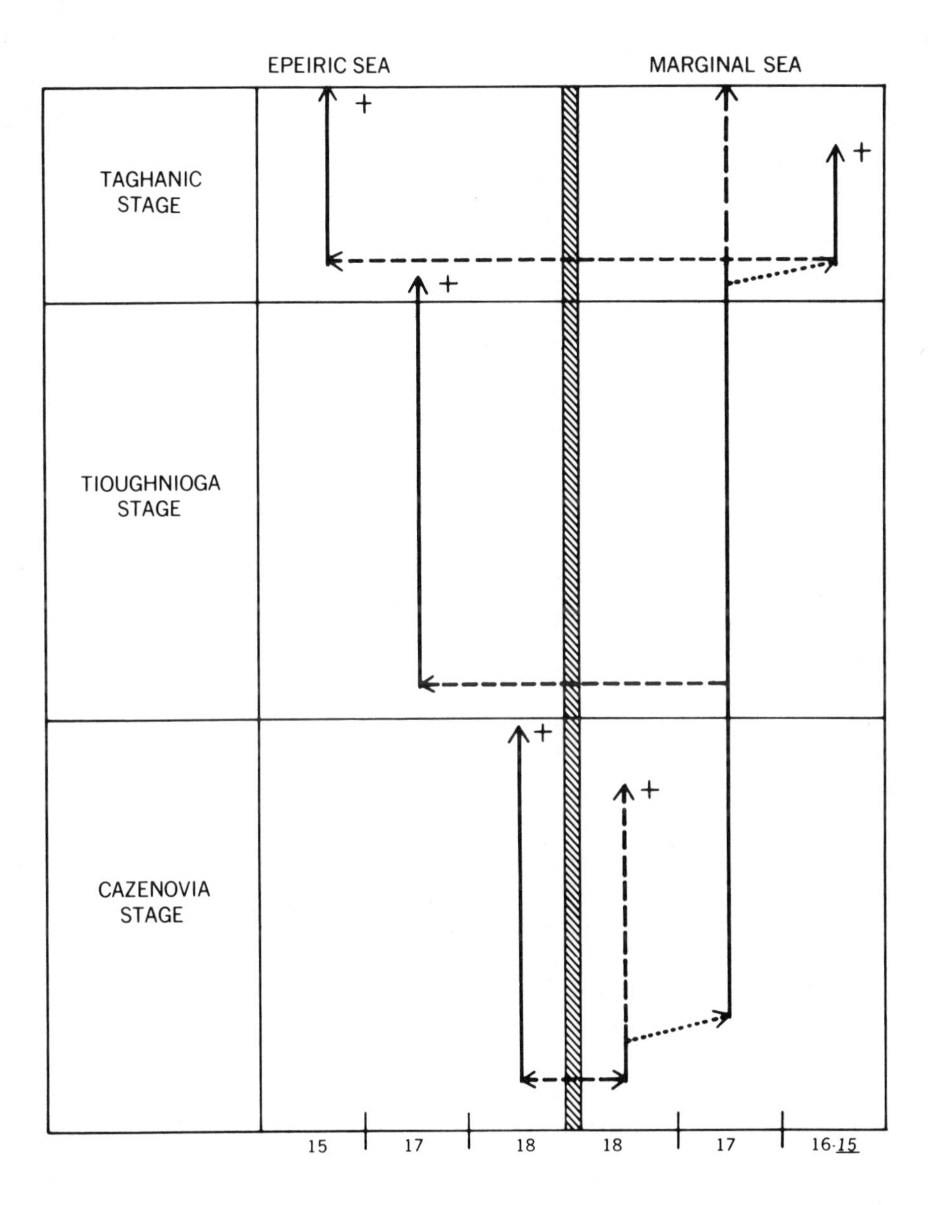

Figure 5–8: Hypothesized phylogeny of the *Phacops rana* stock in the Middle Devonian of North America. Numbers at the base of the diagram refer to the population number of dorso-ventral files. Dotted lines: origin of new (reduced) number of d.-v. files in a peripheral isolate; horizontal dashed lines: migration; vertical solid lines: presence of taxon in indicated area; dashed vertical lines: persistence of ancestral stock in a portion of the marginal sea other than that in which the derived taxon occurs. Crosses denote final disappearance; for fuller explanation, see text.

Our two examples, so widely separated in scale, age, and subject, have much in common as exemplars of allopatric processes. Both required an attention to details of *geographic* distribution for their elucidation. Both involved a *more literal* reading of the fossil record than is allowed under the unconscious guidance of phyletic gradualism. Both are characterized by *rapid* evolutionary events punctuating a history of stasis. These are among the expected consequences if most fossil species arose by allopatric speciation in small, peripherally isolated populations. This alternative picture merely represents the application to the fossil record of the dominant theory of speciation in modern evolutionary thought. We believe that the consequences of this theory are more nearly demonstrated than those of phyletic gradualism by the fossil record of the vast majority of Metazoa.

Some Extrapolations to Macroevolution

Before 1930, paleontology sought a separate theory for the causes of macroevolution. The processes of microevolution (including the origin of species) were deemed insufficient to generate the complexity and diversity of life, even under the generous constraint of geological time; a variety of special causes were proposed—vitalism, orthogenesis, racial "life" cycles, and universal acceleration in development to name just a few.

However, the advent of the "modern synthesis" inspired a reassessment that must stand as the major conceptual advance in 20th-century paleontology. Special explanations for macroevolution were abandoned for a simplifying theory of extrapolation from species-level processes. All evolutionary events, including those that seemed most strongly "directed" and greatly extended in time, were explained as consequences of mutation, recombination, selection, etc.—i.e., as consequences only of the phenomena that produce evolution in nature's real taxon, the species. (The modern synthesis received its name because it gathered under one theory—with population genetics at its core—the events in many subfields that had previously been explained by special theories unique to that discipline. Such an occurrence marks scientific "progress" in its truest sense—the replacement of special explanations carrying little power in prediction or extension with general theories, rich in implications and capable of unifying a diverse set of phenomena that had seemed unrelated. Thus Simpson (1944, 1953) did for paleontology what Dobzhansky (1937) had done for classical genetics, Mayr (1942) for systematics, de Beer (1940) for development, White (1954) for cytology, and Stebbins (1950) for systematic botany—he exemplified the phenomena of his field as the result of Darwinian processes acting upon species.)

We have discussed two pictures for the origin of species in paleontology. In the perspective of a species-extrapolation theory of macroevolution, we should now extend these pictures to see how macroevolution proceeds under their guidance. If actual events, as recorded by fossils, fit more comfortably with the predictions of either picture, this will be a further argument for that picture's greater adequacy.

Under phyletic gradualism, the history of life should be one of *stately unfolding*. Most changes occur slowly and evenly by phyletic transformation; splitting, when it occurs, produces a slow and very gradual divergence of forms (Weller's (1969) tree of life—reproduced as *figure 5–9*—records the extrapolation of this partisan view, not a neutral hatrack for the fossils themselves). We have already named our alternate picture for its predicted extrapolation—*punctuated equilibria*. The theory of allopatric speciation implies

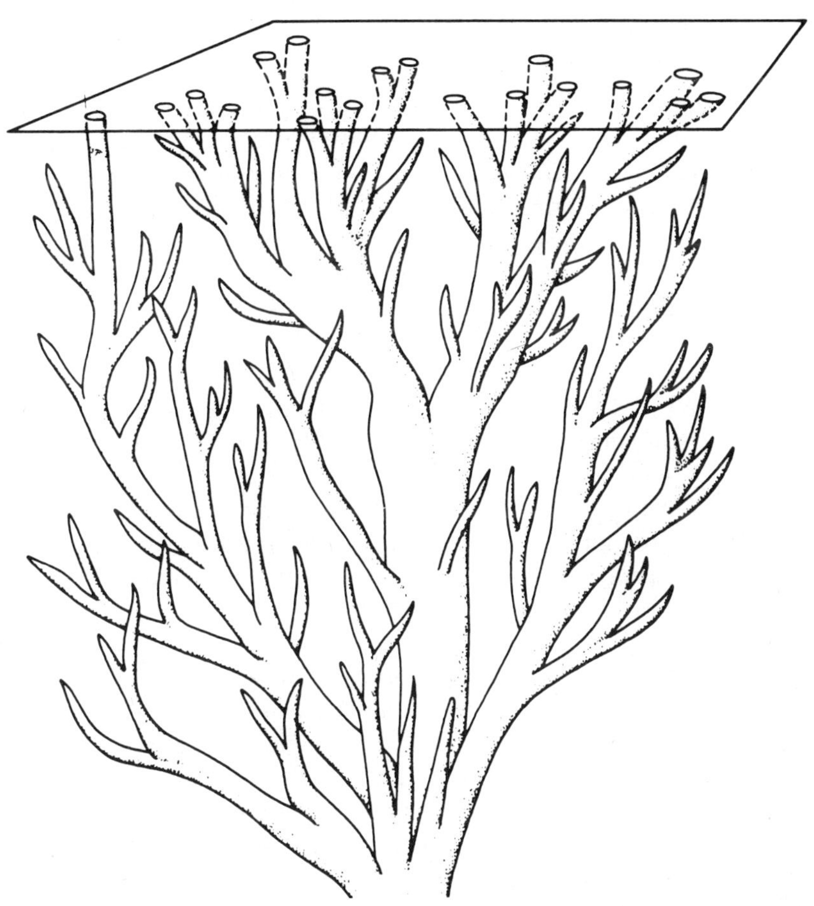

Figure 5–9: The "Tree of Life" viewed from the perspective of phyletic gradualism. Branches diverge gradually one from the other. A slow and relatively equal rate of evolution pervades the system. From Weller, 1969; figure 637.

that a lineage's history includes long periods of morphologic stability, punctuated here and there by rapid events of speciation in isolated subpopulations.

We now consider two phenomena of macroevolution as case studies of our extrapolated pictures. The first is widely recognized as anomalous under the unconscious guidance of stately unfolding; it emerges as an expectation under the notion of punctuated equilibria. The second phenomenon seems, superficially, to have an easier explanation under stately unfolding, but we shall argue that it has a more interesting interpretation when viewed with the picture of punctuated equilibria.

(1) *"Classes" of great number and low diversity*

To many paleontologists, nothing is more distressing than the current situation in echinoderm systematics. Ubaghs (1967), in his contribution to the *Treatise on Invertebrate Paleontology*, recognizes 20 classes and at least one has been added since then—Robison and Sprinkle's (1969) ctenocystoids. Yet, although all appeared by the Ordovician, only five survived the Devonian. Moreover, although each class has a distinct Bauplan, many display a diversity often considered embarrassingly small for so exalted a taxonomic rank—the *Treatise* describes eight classes with five or fewer genera; five of these include but a single genus (as does the new ctenocystoids).

There are two aspects to this tale that fit poorly with the traditional view of stately unfolding:

(1) The presence of 21 classes by the Ordovician, coupled with their presumed monophyletic descent, requires extrapolation to a common ancestor uncomfortably far back in the Precambrian if Ordovician diversity is the apex of a gradual unfolding. Yet current views of Precambrian evolution will not happily accommodate a complex metazoan so early (Cloud, 1968).

(2) We expect that successively higher ranks of the taxonomic hierarchy will contain more and more taxa: a class with one genus is anomalous and we are led either to desperate hopes for synonymy or, once again, to our old assumption—that we possess a fragmentary record of a truly diverse group. Yet this expectation is no consequence of the logic of taxonomy (which demands only that each taxon be *as* inclusive as the lower ones it incorporates); it arises, rather, from a picture of stately unfolding. In *figure 5–9*, a new higher taxon attains its rank *by virtue of* its diversity—an evenly progressing, evenly diverging set of branches cannot produce such a taxon with limited diversity, for a lineage "graduates" from family to order to class only as it persists to a tolerable age and branches an acceptable number of times.

With the picture of punctuated equilibria, however, classes of small membership are welcome and echinoderm evolution becomes more intriguing than bothersome. Since speciation is rapid and episodic, repeated splitting during short intervals is likely when opportunities for full speciation following isolation are good (limited dangers of predation or competition in peripheral environments, for example—a likely Lower Cambrian situation). When these repeated splits affect a small, isolated lineage; when adaptation to peripheral environments involves new modes of feeding, protection, and locomotion; and when extinction of parental species commonly follows the migration of descendants to the ancestral area, then very distinct phenons with few species will develop. Since higher taxa are all "arbitrary" (they reflect no interacting group in nature, but rather a convenient arrangement of species that violates no rule of monophyly, hierarchical ordering, etc.), we believe that they should be defined by morphology. Criteria of diversity are too closely tied to partisan pictures; morphology, though not as "objective" as some numerical taxonomists claim, is at least more functional for information retrieval.

(2) *Trends*

Trends, or biostratigraphic character gradients, are frequently mentioned as basic features of the fossil record. Sequences of fossils, said to display trends, range from the infraspecific through the very highest levels of the taxonomic hierarchy. Trends at and below the species level were discussed in the previous section, but the relation between phyletic gradualism and trends among related clusters of species—families or orders—remains to be examined.

Many, if not most, trends involving higher taxa may simply reflect a selective rendering of elements in the fossil record, chosen because they seem to form a morphologically-graded series coincident with a progressive biostratigraphic distribution. In this sense, trends may represent simple extrapolations of phyletic gradualism.

But a claim that all documented trends are just unwarranted extrapolations based on a preconception would be altogether too facile an explanation for the large numbers of trends cited in the literature. For this discussion, we accept trends as a real and important phenomenon in evolution, and adopt the simple definition given by MacGillavry (1968, p. 72): "A trend is a direction which involves the *majority* of related lineages of a group" (our italics).

If trends are real and common, how can they be reconciled with our picture, in which speciation occurs in peripheral isolates by adaptation to local conditions and the perfection of isolating mechanisms? The problem may be stated in another way: Sewall Wright (1967, p. 120) has suggested that, just as mutations are stochastic with respect to selection within a population, so might speciation be stochastic with respect to the origin of higher taxa. As a slight extension of that statement, we might claim that adaptations to local conditions by peripheral isolates are stochastic with respect to long-term, net directional change (trends) within a higher taxon as a whole. We are left with a bit of a paradox: to picture speciation as an allopatric phenomenon, involving rapid differentiation within a general, long-term picture of stasis, is to deny the picture of directed gradualism in speciation. Yet, superficially at least, this directed gradualism is easier to reconcile with valid cases of long-term trends involving many species.

MacGillavry's definition of a trend removes part of the problem by using the expression "majority of related lineages." This frees us from the constraint of reconciling *all* events of adaptation to local conditions in peripheral isolates, with long-term, net directional change.

A reconciliation of allopatric speciation with long-term trends can be formulated along the following lines: we envision multiple "explorations" or "experimentations" (see Schaeffer, 1965)—i.e., invasions, on a stochastic basis, of new environments by peripheral isolates. There is nothing inherently directional about these invasions. However, a subset of these new environments might, in the context of inherited genetic constitution in the ancestral components of a lineage, lead to new and improved efficiency. Improvement would be consistently greater within this hypothetical subset of local conditions that a population might invade. The overall effect would then be one of net, apparently directional change: but, as in the case of selection upon mutations, the initial variations would be stochastic with respect to this change (*figure 5–10*). We postulate no "new" type of selection. We simply state a view of long-term, superficially "directed" phenomena that is in accord with the theory of allopatric speciation, and also avoids the largely untestable concept of orthoselection.

Conclusion: Evolution, Stately or Episodic?

Heretofore, we have spoken of the morphological stability of species in time without examining the reasons for it. The standard definition of a biospecies—as a group of actually or potentially reproducing organisms sharing a common gene pool—specifies the major reason usually cited: gene flow. Since the subpopulations of a species adapt to a range of differing local environments, we might expect these groups to differentiate, acquire isolating mechanisms and, eventually, to form new species. But gene flow exerts a homogenizing influence "to counteract local ecotypic adaptation by breaking up well-integrated gene complexes" (Mayr, 1963, p. 178). The role of gene flow is recognized in the central tenet of allopatric speciation: speciation occurs in *peripheral* isolates because only geographic separation from the parental species can reduce gene flow sufficiently to allow local differentiation to proceed to full speciation.

Recently, however, a serious challenge to the importance of gene flow in species' cohesion has come from several sources (Ehrlich and Raven, 1969, for example). Critics claim that, in most cases, gene flow is simply too restricted to exert a homogenizing influence and prevent differentiation. This

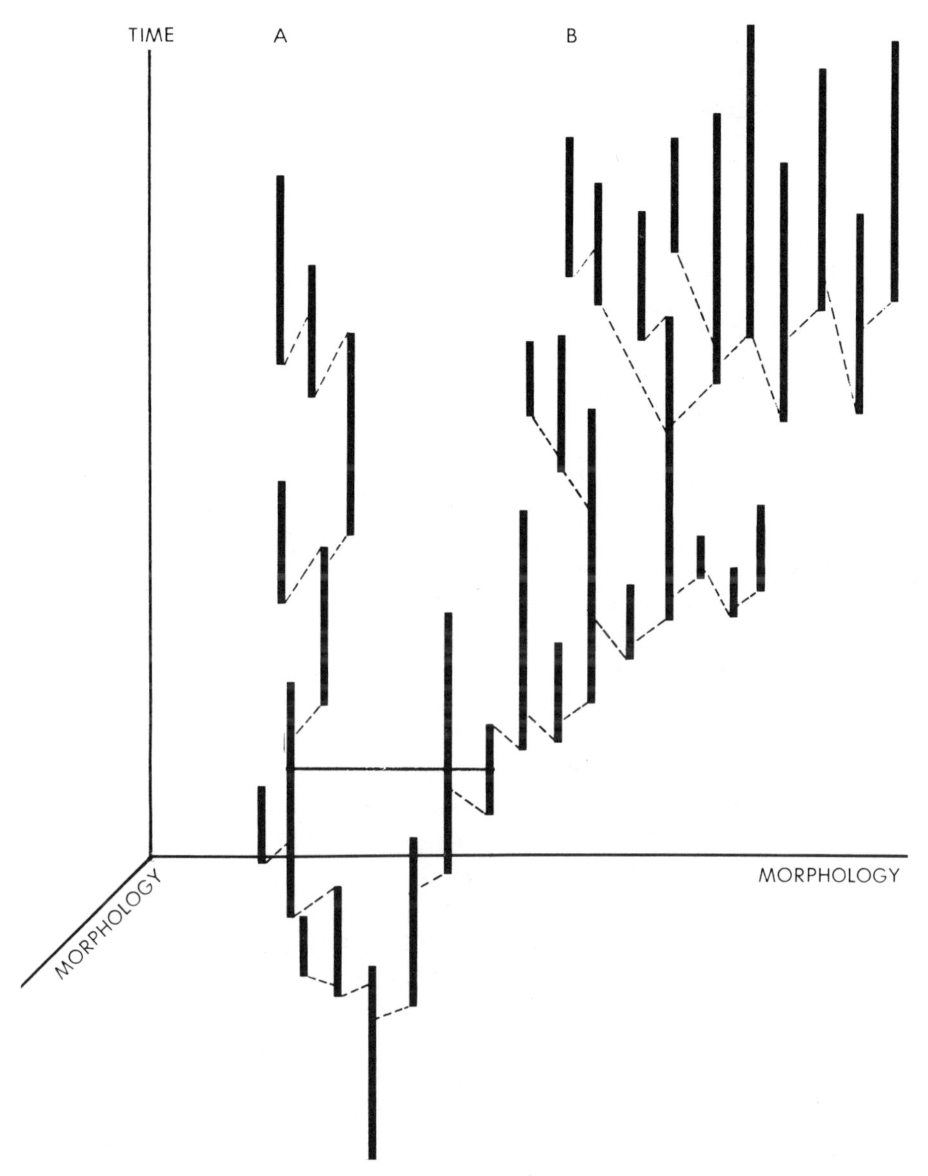

Figure 5–10: Three-dimensional sketch contrasting a pattern of relative stability (A) with a trend (B), where speciation (dashed lines) is occurring in both major lineages. Morphological change is depicted here along the horizontal axes, while the vertical axis is time. Though a retrospective pattern of directional selection might be fitted as a straight line in (B), the actual pattern is stasis within species, and differential success of species exhibiting morphological change in a particular direction. For further explanation, see text.

produces a paradox: why, then, are species coherent (or even recognizable)? Why do groups of (relatively independent) local populations continue to display a fairly consistent phenotype that permits their recognition as a species? Why does reproductive isolation not arise in every local population? Why is the local population itself not considered the "real" unit in evolution (as some would prefer—Sokal and Crovello, 1970, p. 151, for example)?

The answer probably lies in a view of species and individuals as homeostatic systems—as amazingly well-buffered to resist change and maintain stability in the face of disturbing influences. This concept has been urged particularly by Lerner (1954) and Mayr (1963), though the latter still gives more weight to gene flow than many will allow. Lerner (1954, p. 6) recognizes two types of homeostasis, mediated in both cases, he believes, by the generally higher fitness of heterozygous vs. homozygous genotypes: (1) ontogenetic self-regulation (developmental homeostasis) "based on the greater ability of the heterozygote to stay within the norms of canalized development" and (2) self-regulation of populations (genetic homeostasis) "based on natural selection favoring intermediate rather than extreme phenotypes." In this view, the importance of peripheral isolates lies in their small size and the alien environment beyond the species border that they inhabit—for only here are selective pressures strong enough and the inertia of large numbers sufficiently reduced to produce the "genetic revolution" (Mayr, 1963, p. 533) that overcomes homeostasis. The coherence of a species, therefore, is not maintained by interaction among its members (gene flow). It emerges, rather, as an historical consequence of the species' origin as a peripherally isolated population that acquired its own powerful homeostatic system. (We regard this idea as a serious challenge to the conventional view of species' reality that depends upon the organization of species as ecological units of *interacting* individuals in nature. If groups of nearly-independent local populations are recognized as species only because they share a set of homeostatic mechanisms developed long ago in a peripheral isolate that was "real" in our conventional sense of interaction, then some persistent anomalies are resolved. The arrangement of many asexual groups into good phenetic "species," quite inexplicable if interaction is the basis for coherence, receives a comfortable explanation under notions of homeostasis.)

Thus, the challenge to gene flow that seemed to question the stability of species in time ends by reinforcing that stability even more strongly. If we view a species as a set of subpopulations, all ready and able to differentiate but held in check only by the rein of gene flow, then the stability of species is a tenuous thing indeed. But if that stability is an inherent property both of individual development and the genetic structure of populations, then its power is immeasurably enhanced, for the basic property of homeostatic systems, or steady states, is that they resist change by self-regulation. That local populations do not differentiate into species, even though no external bar prevents it, stands as strong testimony to the inherent stability of species in time.

Paleontologists should recognize that much of their thought is conditioned by a peculiar perspective that they must bring to the study of life: they must look down from its present complexity and diversity into the past; their view must be retrospective. From this vantage point, it is very difficult to view evolution as anything but an easy and inevitable result of mere existence, as something that unfolds in a natural and orderly fashion. Yet we urge a different view. The norm for a species or, by extension, a community is stability. Speciation is a rare and difficult event that punctuates a system in homeostatic equilibrium. That so uncommon an event should have produced such a wondrous array of living and fossil forms can only give strength to an old idea: paleontology deals with a phenomenon that belongs to it alone among the evolutionary sciences and that enlightens all its conclusions—time.

6

MODELS FOR THE EVOLUTION OF PLANKTONIC FORAMINIFERA

Francis G. Stehli · Robert G. Douglas · Ismail A. Kafescioglu

---

Editorial introduction. The range of faunal and floral diversity varies over the surface of the earth, and the causes for this are currently of very intense interest. By *choosing* a group of animals about which sufficient information is known (in this case, planktonic foraminifera), Stehli, Douglas, and Kafescioglu develop and test a model which predicts that the average evolutionary rate of species is independent of latitude. This important prediction is confirmed. Variation in the range of diversity is explained by supposing that there are more species in some areas (tropics) than in others (high latitudes) because of greater opportunities (more unclaimed niches) for species to arise.

Two important and more general points emerge from their essay. First, paleontology at the present time is in the phase of development in which models based on living organisms are being proposed. Secondly, the authors show beautifully the steps in constructing an argument: the explicit enumeration of the model, the test, and adjustment in the model. In this respect, this paper could well serve as a guide for students.

Introduction

Ideally the fossil record should provide a testing ground for hypotheses concerned with the mechanisms of evolution. Unfortunately, its utility for this purpose is marred both by actual imperfections in the record and by limitations in our knowledge of it, and it is, therefore, seldom adequate to allow definitive testing of fine detail. Using a combination of biological and geological data Stehli and Wells (1971) were able to demonstrate a general correlation between mean annual temperature and apparent evolutionary rates for hermatypic coral genera. Evolution, however, proceeds at the level of the species rather than the genus and it would be desirable to carry examination of the relationship of temperature and evolutionary rates to this level. Neither the taxonomy nor the geographic distribution of species of hermatypic corals is sufficiently well known to allow this. It now appears, however, that for the planktonic foraminifera the fossil record and Recent distribution may be sufficiently described to permit consideration of the relationship between temperature and evolutionary rate at the species level. Even with this relatively well known and intensively studied group, it is necessary initially to be content with the use of averages, and to accept as "the state of the art" some taxonomic system even though it is recognized that our understanding of the group is growing apace, and that any system is therefore ephemeral. We propose to use data on the diversity and average age of planktonic foraminifera at both the generic and the specific level to test the hypotheses that evolution is proceeding more rapidly in the warmer portions of the globe than elsewhere.

A Model

Stehli and Wells (1971) have established that evolutionary rates in genera at least among hermatypic corals are correlated with temperature. It appears reasonable to assume that genera, being composites of species, should reflect the environmental controls imposed on evolution at the species level. For this reason we propose and plan to test a model which states:

> *Both genera and species of planktonic foraminifera are influenced by temperature in such a way that evolutionary rates are highest in the regions of highest mean annual sea surface temperature. The relative rates of evolution are shown by levels of diversity, which are highest where evolution is most rapid, and by average generic and specific ages, which are lowest where evolution is most rapid.*

The proposed model can be tested by examining biological and geological data on planktonic foraminifera to see whether the predicted behavior is found to exist for both genera and species. Such a test will require, for both taxonomic levels, diversity data for a latitudinally long traverse and concomitant temperature data. Calculation of average age data requires that the range in time of living planktonic foraminiferal genera and species is reasonably well known.

Available Data

Basic distribution. Boltovskoy (1969) published a detailed study of collections made on a traverse along meridian 90°E between the equator and 65°S by the British Royal Research ship *Discovery II* during the year 1951 (*figure 6–1*). At each station along this traverse, plankton hauls were made through

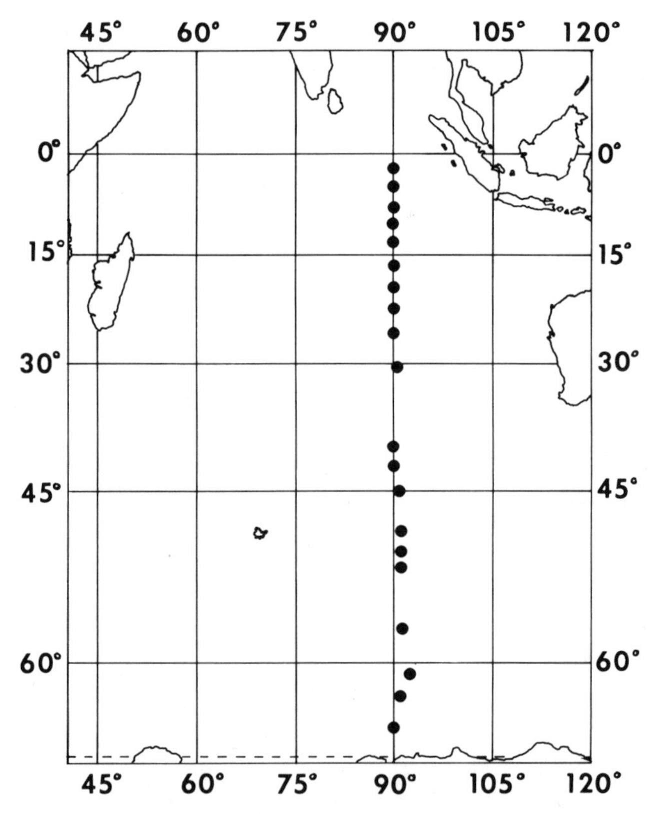

Figure 6–1: Sample stations for the 1951 collections made by the Royal Research Ship *Discovery II* from which Boltovskoy (1969) studied the planktonic foraminifera.

the depth range 0–1500 m and the foraminiferal species present identified. The same kind of sampling equipment was used throughout, yielding data of good internal consistency, but unfortunately the mesh of the plankton nets used was large (0.24 mm) and according to Boltovskoy (1969) some smaller species of planktonic foraminifera were caught only when they accidentally

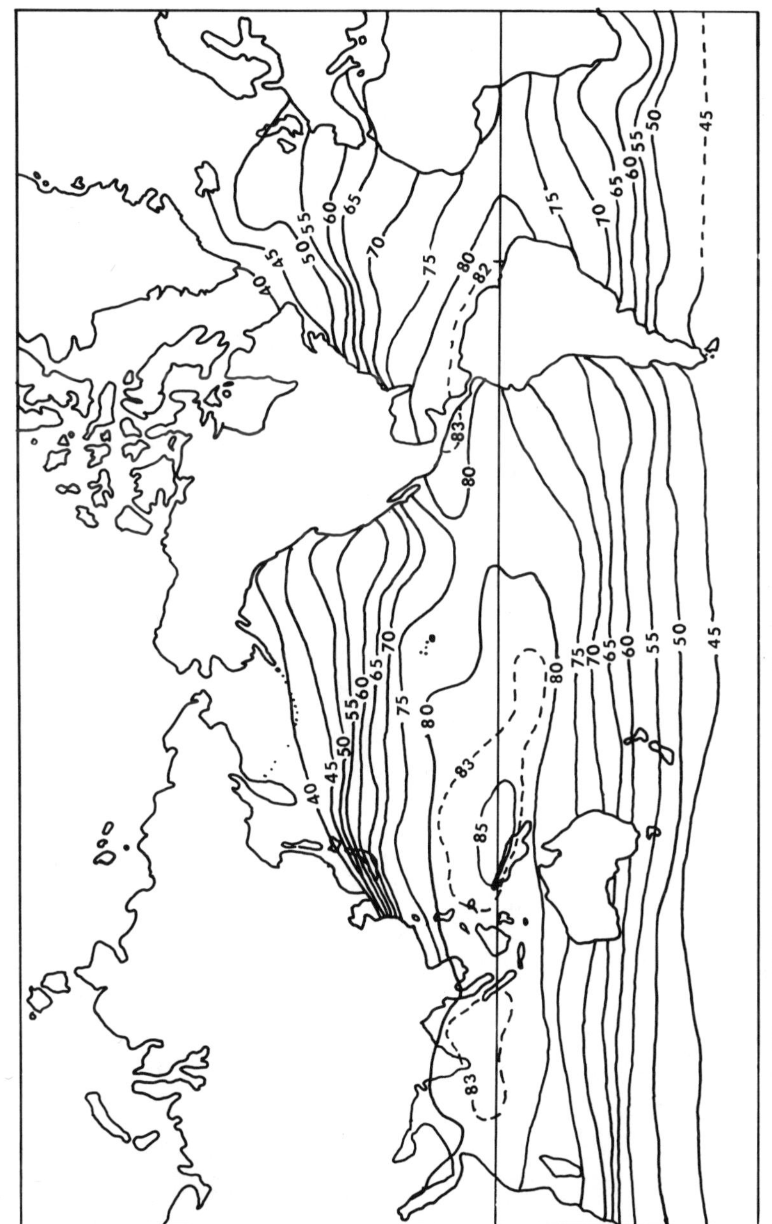

Figure 6–2: Mean annual sea surface temperature (°F) calculated from the monthly averages in the U.S. Navy Marine Climatic Atlas on the World (1955–1963).

became entangled with other planktonic organisms. It seems likely that this suite of data is prejudiced in favor of larger species and against smaller ones, but the prejudice would appear to be systematic and affect all samples, so that it can be ignored in the present study unless a systematic size gradient exists across latitude, which we do not believe to be the case. The suite of samples taken in 1951 is believed by Boltovskoy to have been stored in solutions adequately buffered to prevent solution loss of the more delicate and thinner shelled species. Belyaeva (1963, 1964) has also studied the distribution of planktonic foraminifera from plankton tows and sediment samples in the Indian Ocean. Her data plus unpublished plankton data available to us provided a check of Boltovskoy's results and confirmed that the inferred

Table 6–1. Taxonomic system used in this report and its relationship to that employed by Boltovskoy (1969). If no entry occurs in the second column, then our usage is the same as that of Boltovskoy. The ages of the genera and species as presented by Berggren (1969), Blow (1969), and as used in

Taxonomy employed in this study	Taxonomy employed by Boltovskoy (1969) if different from this report	Age (in 10^6 year) Berggren 1969	Blow 1969	This report
Candeina		11.5	11.5	12
nitida			9	9
Globigerina				62
bulloides		10	11	11
pachyderma		3.5	3.5	4
quinqueloba			9	9
rubescens			3.5	4
Globigerinita				46
glutinata			24	24
uvula	*G. bradyi*		26	26
Globigerinoides			26	26
conglobatus		7	9	9
ruber		10	11	11
sacculifer			23	23
trilobus			23	23
Globoquadrina		30	28	28
conglomerata	*Globigerina conglomerata*		5	5
dutertrei	*Globigerina dutertrei*	3.5	3.5	4
hexagona	*Globigerina hexagona*		7	7

species latitudinal ranges given by Boltovskoy (1969) are reasonably correct. Our analysis is, therefore, confined to the 1951 data. *Table 6–1* lists the species employed in the study.

Age data. By employing information on geologic ranges for genera and species of planktonic foraminifera drawn from Blow (1969) and calibrating the relative time scale in millions of years (Berggren, 1969), it becomes possible to assign an average age for genera and species encountered in each assemblage (*table 6–1*).

Temperature data. Data for mean annual sea surface temperature were calculated from monthly averages given in the U.S. Navy Marine Climatic Atlas of the World (1955–1963). *Figure 6–2* presents the resulting temperature surface for comparison with diversity and average age data.

(*Table 6-1, cont.*)

this report. *Globorotalia cavernula* has been omitted from the age calculations because we were unable to determine its geologic range to our satisfaction.

Globorotalia		26	23	26
crassaformis		10	11	10
culturata	*G. menardii*	16.5	14	14
hirsuta			2	2
pumilia				2
scitula			18	18
truncatulinoides		1.85	1.85	2
cavernula		—	—	—
tumida	included as a variety of *G. menardii*	7	7	7
Turborotalia				10
inflata	*Globoratalia inflata*	8	9	9
Hastigerinella				
adamsi	*Globigerinella adamsi*		0.5	1
Hastigerina			21	21
pelagica			11	11
siphonifera	*Globigerinella aeguilateralis*	16	16	16
Orbulina		18.5	18	18
universa		18	18	18
Pulleniatina		8	8	8
obliguiloculata		4	5	5
Sphaeroidinella		5	5	5
dehiscens	included in *Globigerinoides trilobus*	5	5	5

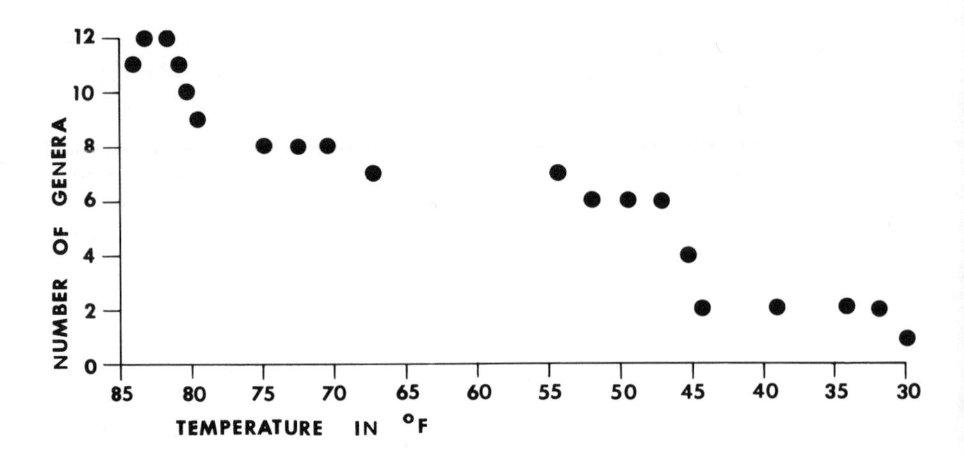

Figure 6–3: Generic diversity of planktonic foraminifera as derived from Boltovskoy (1969) with minor additions and plotted against mean annual sea surface temperature (°F). Temperatures below 45°F were taken from a 2nd order non-orthogonal polynomial surface calculated to fit the data of *figure 6–2*.

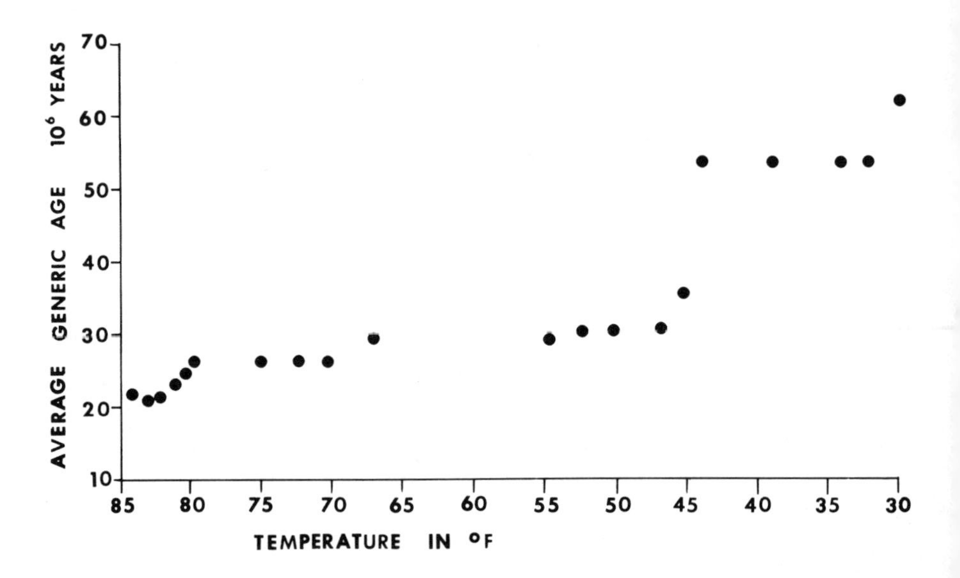

Figure 6–4: Average generic age for planktonic foraminifera as derived from the lists of Boltovskoy (1969) with minor additions and plotted against mean annual sea surface temperature (°F). Temperatures below 45°F were taken from a 2nd order non-orthogonal polynomial surface calculated to fit the data of *figure 6–2*.

Generic diversity data. The generic diversity at the stations shown in *figure 6–1*, normalized to percent of the total number of genera recognized, is shown plotted against mean annual sea surface temperature in *figure 6–3*. It is seen that the generic diversity is a relatively smooth function of temperature, but that important temperature thresholds are present. If high diversity is related to rapid evolution, then the data seem to support the proposed model.

Average generic age. The average generic age for each sample station has been determined by dividing the number of genera represented into the combined ages of the genera present expressed in millions of years. The measure is somewhat crude because the relative time ranges available for calibration are often no more definitive than "Upper Eocene to Recent." *Figure 6–4* shows a plot of the average generic age found for each station against mean annual sea surface temperature. The relationship seen is that predicted by the model and one must conclude that if survivorship is indeed the reciprocal of evolutionary rate, as Simpson (1953) suggested, then genera of planktonic foraminifera suggest a correlation between increasing temperature and increasing evolutionary rates, modified somewhat by temperature thresholds.

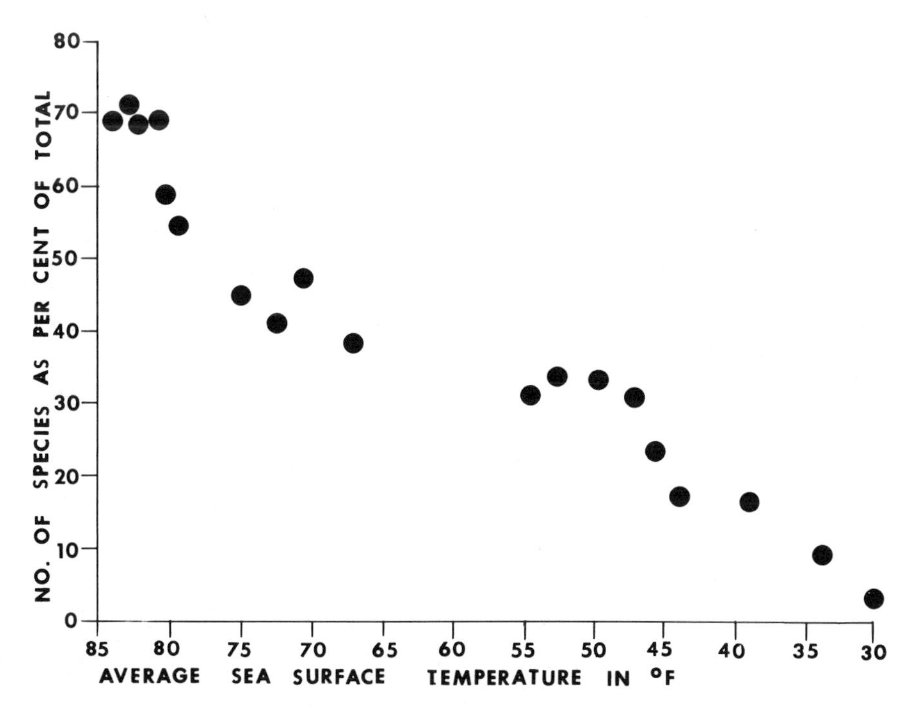

Figure 6–5: Specific diversity of planktonic foraminifera as derived from Boltovskoy (1969) with minor additions and plotted against mean annual sea surface temperatures (°F). Temperatures below 45°F were taken from a 2nd order non-orthogonal polynomial surface calculated to fit the data of *figure 6–2*.

Species diversity. The species level diversity at the stations of *figure 6–1*, normalized to percent of the total number of species recognized by Boltovskoy, is shown plotted against mean annual sea surface temperature in *figure 6–5*. This curve can be compared to that for genera (*figure 6–3*) and it will be noted that both curves have the same form. Thus, in the case of diversity, it appears that our model is correct and that, insofar as high diversity is related to high evolutionary rates, these rates are correlated with temperature at both taxonomic levels.

Average specific age. The average specific age at each of the stations can be calculated in a manner analogous to that used for the average generic age. Once more the measure is somewhat crude because of some arbitrariness in assigning a particular number of millions of years to represent what is actually a range of time, such as Middle Miocene to Recent. In *figure 6–6* the average specific age is plotted against mean annual sea surface temperature. The average species age is found to differ strikingly from the average generic age plot in that species show only a very slight relationship to temperature. This unexpected result is not predicted by our initial model, which is thus shown to be incorrect, and in need of modification.

Analysis of Data

At the level of species we have data for diversity distribution and for average age. The diversity data indicate that the number of species increases as a function of increasing temperature. The average age curve shows us that, regardless of latitude, the evolution of species occurs *on the average* at a nearly constant rate. This fact makes it apparent that the observed diversity curve is unlikely to result from a higher absolute rate of evolution in the warmer parts of the temperature spectrum. We may tentatively conclude, then, that it must result either from coincidence or from some temperature dependence in the availability of unclaimed niches within which evolution can take place. There is sufficient documentation of the fact that temperature dependence in diversity is the normal pattern (Fischer, 1960; Stehli, 1965; Stehli, 1968) to foreclose the possibility of coincidence as the cause of this pattern in species of planktonic foraminifera. If coincidence is not the cause, and average evolutionary rates are nearly constant for species as a function of temperature, then we must conclude that the observed diversity gradient owes its form to more rapid net production of species in the warmer portions of the temperature spectrum. This phenomenon could occur even with constant average evolutionary rates if there were in warmer regions a greater availability of unclaimed niches within which evolution of species could occur.

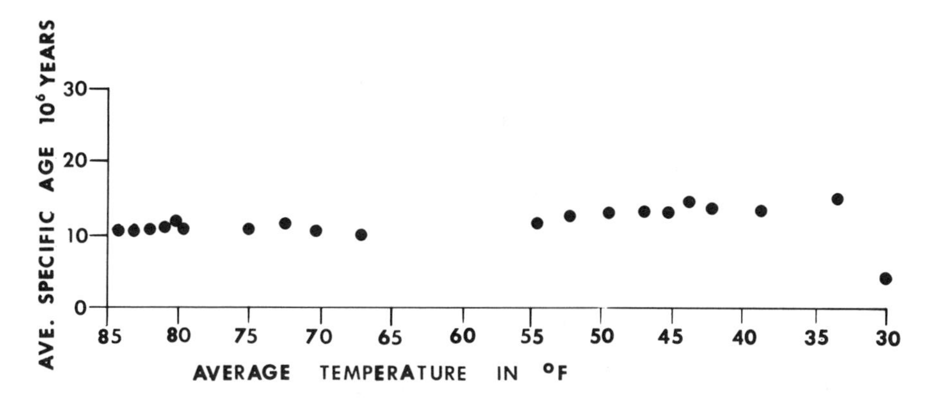

Figure 6–6: Average specific age for planktonic foraminifera as derived from the lists of Boltovskoy (1969) with minor additions plotted against mean annual sea surface temperatures (°F). Temperatures below 45°F were taken from a 2nd order non-orthogonal polynomial surface calculated to fit the data of *figure 6–2.*

At the level of genus we once more have both diversity distribution data and average age data. The diversity curve for genera is very similar to that found for species and is once again so similar to the pattern characteristic of other large groups of easily distributed organisms that we can eliminate coincidence as a possible explanation of temperature dependence. Genera, of course, are abstractions recognized only as a convenient means of dealing with groups of closely related species and as such do not have "evolutionary rates" in the same sense as species. The average "evolutionary rate" of a genus, it would appear, must depend on the average evolutionary rate of its component species. We see, thus, that average "evolutionary rates" for genera cannot be expected to differ qualitatively from average evolutionary rates for species, and we see also that the form of the diversity curve at the two levels is the same. From these data we may conclude that genera too owe their higher net rate of production in regions of higher temperature to some temperature dependence in the availability of unclaimed niches. In the case of genera, however, the kind of niche involved must be one with sufficient potential to allow secondary speciation by the occupying species and the temperature dependence probably is to be found in the availability of unclaimed niches of high potential.

Clearly, the average generic age curve indicates the existence of some factor operative in production of genera that does not affect, to nearly as large a degree, the production of species. It is of interest to attempt to isolate and identify this factor. The observed patterns in generic distribution and age indicate the presence of large numbers of young genera in the tropics and the presence of a few, generally old, genera in cold-water regions. From the available data one tentatively can conclude that new adaptive types (genera) originate mainly through tropical species and then through secondary speciation spread into less favorable extra-tropical areas. The existence of temperature thresholds in the curves of figures 6–3 through 6–5 suggests that water masses between thresholds are colonized rapidly (instantly from a geologic point of view). The development of more tolerant species must then occur before less favorable (colder) water masses are occupied by the genus (in preparation, Douglas and Stehli).

An Alternative Model

The observed data show us that our initial model was incorrect and requires revision. Our analysis of the data suggests a new model which states:

The average evolutionary rate of species of planktonic foraminifera does not vary greatly with temperature, but the number of species evolved (diversity) does vary with temperature because it influences the number of unclaimed niches available for invasion by new species. Strong temperature control over the ease of access to unclaimed niches of evolutionary potential high enough to allow secondary speciation (genus formation) causes most new genera to appear in the warm-water region (figure 6–7). Invasion of regions in which the newly occupied niche is represented proceeds step-wise across a temperature gradient so that genera are old when their species occupy the extremities of the possible temperature range.

Test of Alternative Model

If the revised model is correct, one would expect genera still restricted to tropical regions to be young and to tend toward monotypy, while genera occurring in marginal areas should be old, cosmopolitan and polytypic. This prediction can be used as a test of the validity of the model if appropriate data on the distribution of monotypic and polytypic genera can be obtained. From the basic data used earlier, one can draw information regarding all the necessary parameters, age, number of species per genus and distribution; thus, a preliminary, crude test of the new model is possible. In the northern-most four stations of Boltovskoy's data the average number of species recognized in the genera present is 1.7, while in the southernmost stations the genera present have an average of 2.5 species per genus. In the single southern-most station at 65°S only the venerable genus *Globigerina* is present. This genus has six species in the Indian Ocean, and it is more than 62 million years old and cosmopolitan. In the equatorial northern end of the traverse, two

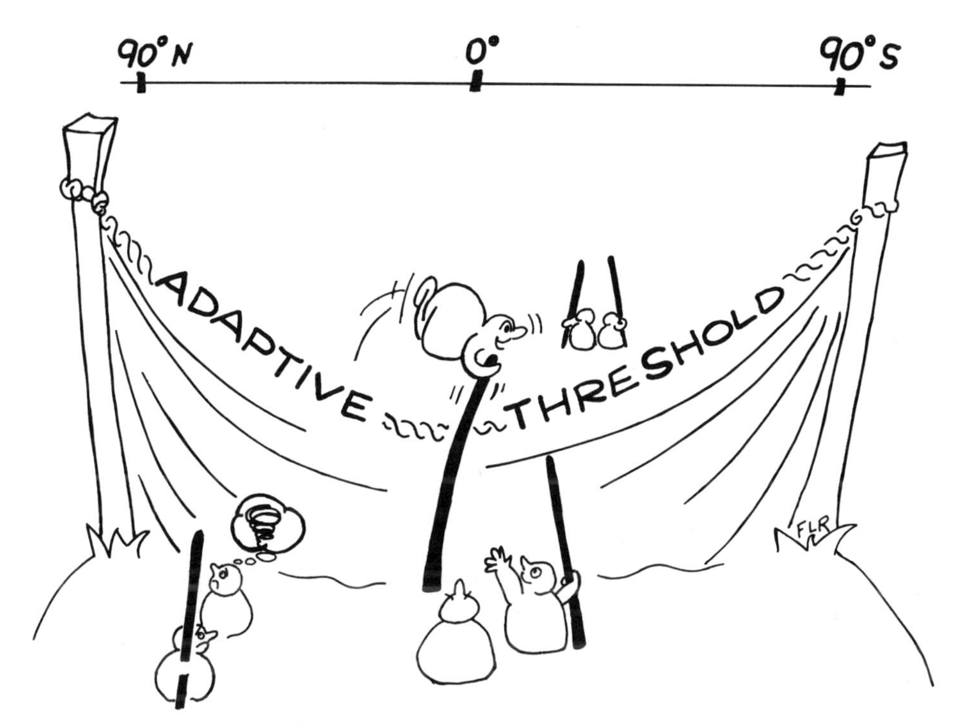

Figure 6–7: A simplified model of our concept of the development of new genera principally in the tropics. The barrier labelled *Adaptive Threshold* is that beyond which lie unclaimed niches of sufficient potential to allow within them the secondary speciation that produces new genera. Note that these unclaimed niches exist at all temperatures but that the barrier to admission becomes more difficult to cross as temperature decreases along a traverse. Threshold crossing species are most likely to be successful in crossing the barrier in the tropics when it is low, but once across they can extend occupation of their niche slowly down the temperature gradient by means of secondary speciation. The development of secondary species to fill all accessible portions of the generic niche proceeds at rates not strongly dependent on temperature.

genera, *Pulleniatina* and *Candeina*, occur which do not persist beyond 15°S. (Berggren, personal communication, 1971, suggests that there is some difference of opinion as to the monotypy of *Pulleniatina*. We have assumed a taxonomic system for purposes of analysis and it is arbitrary and subject to debate. A general increase in number of species in genera with broad latitudinal ranges over those with warm-water distributions is nevertheless quite clear.) Both of these genera are monotypic, highly restricted to the tropics and young (average age of 8 million years). This test is crude but is consistent with the revised model.

Additional support for the model would be furnished if it could be demonstrated that genera now widespread were formerly restricted to the tropics. This kind of data is extremely difficult to get because the strata from which planktonic genera are obtained tend to be assigned ages on the basis of the planktonic genera present. Only if geochemical age dates were available at each station for strata exhibiting the first occurrence of the representatives of some genus across a considerable range of latitudes would it be possible to further substantiate the model. Such data do not exist as far as we are aware, and a conclusive test of the alternative model is not yet possible.

More quantitative data on the time and place of origin of genera of planktonic foraminifera are certainly to be desired, but at the present time the model of virtually exclusive origin of genera of planktonic foraminifera from tropical species entering unclaimed niches of high potential and then spreading into higher latitude, less favorable, and generally cooler, areas in a step-wise fashion by secondary speciation seems to furnish a reasonable working hypothesis.

If the hypothesis is correct, it has extremely important consequences for geology because it suggests that attempts at paleontological correlation across significant ranges of latitude or across temperature gradients, using the first occurrence of genera or higher taxa as biostratigraphic "time planes," will inevitably result in time transgressive surfaces. If such a phenomenon is widespread, and the paleobotanical evidence of Smiley (1969) and coral data of Stehli and Wells (1971) suggest that it may be, it may introduce, especially in the Cenozoic, a large percentage error in correlations and have important implications for biostratigraphy and geologic history. It appears to be very important to get as quickly as possible both ranges of planktonic foraminifera and quantitative age determinations across a large range of latitudes.

7

MODELS IN PHYLOGENY

Michael T. Ghiselin

Editorial introduction. Phylogeny has long been a major goal of some paleontologists. However, both methodologically and philosophically, the preparation of phylogenetic studies is very subtle and difficult—to do correctly. As one who has been concerned with several real-world problems, but who has also taken special pains to consider the philosophical implications of his methodology, Ghiselin discusses models in phylogeny. The fact that philosophical precepts greatly influence, and even dictate, the way in which studies are carried out is developed from another perspective.

Bccausc of its philosophical implications, this chapter can be read on two different levels. As one reviewer stated: "The remarkable thing about most of us is that we use correct reasoning instinctively. Thus, when a philosopher tries to make this reasoning explicit, we are likely to react by calling his work obvious or trivial. Yet we never make the explicit statements ourselves (and are scarcely aware of them until we read the "obvious" systematization of others); moreover, even if we are explicit on given matters, we surely don't string obvious statements together into a coherent general statement. Thus I, for one, strongly support this attempt to piece together our common methodology. I'm also pleased to note that it is a rousing support for paleontology's role in phylogenizing."

Introduction

The term "phylogeny" means the evolutionary development of lineages of organisms. Originally coined by the 19th-century evolutionist Ernst Haeckel, it was contrasted with "ontogeny," the embryological development of individual organisms. The study of phylogeny of course relates to the whole scope of historical biology, whether approached from a paleontological or a neontological point of view. Yet traditionally persons who have concerned themselves with it have emphasized two main activities: reconstructing common ancestors, and drawing "phylogenetic trees." The latter "genealogical" aspect is very important for systematics, although opinion differs as to how much a classification system can, or should, reflect branching sequences (see Hull, 1964).

Phylogeny, indeed, is a most controversial subject. A large body of related literature even purports to debunk the whole field. Characteristically, these attacks have been written by persons untrained in comparative anatomy who maintain, not upon the basis of what has actually been done, but upon questionable philosophical grounds, that the goals of the subject are unattainable. Nonetheless it is true that many works on phylogeny do read like imaginative literature rather than science. A disproportionate segment of the literature seeks to fill gaps in the data with speculations and nothing more. Yet after all, we should expect the greatest interest to focus upon doubtful matters. One receives little credit for working on problems that have already been solved. Hence we must reject as sophistical those critiques which point to uncertainties and ignore many successes.

Thus much of the difficulty in knowing phylogeny may be attributed to insufficient data. But that is by no means the whole problem. There seems to be very little consensus as to how one does phylogenetical research. And different methodologies and theoretical assumptions lead to divergent interpretations of the same data. The situation is all the more difficult because in research papers many of the basic premises are not stated explicitly, so that one must read between the lines to understand the authors' points of view. Quite a number of publications have been devoted to phylogenetical methodology (e.g., Davis, 1949; Harrison and Weiner, 1963; Hertwig, 1914; Inger, 1958; Lam, 1936; Maslin, 1952; Zangerl, 1948; Zimmermann, 1959). Yet even here, we find widely differing positions. Many of the most important procedures are hardly even mentioned, and certain schools deliberately exclude one or another kind of evidence. Fine textbooks on systematics have been written by Simpson (1961) and Mayr (1969), but a student who consults them for information about how to do phylogenetical research will find very little to work with. The widely cited books by Remane (1952) and Hennig (1950) contain many interesting ideas, but they represent philosophical points of view to which many evolutionary biologists are strongly opposed

(Mayr, 1965b). Indeed, there exists no single, modern, and comprehensive treatment of the subject in which even the basic operations are explained. For this reason it seems appropriate that we reexamine the fundamentals, bringing the underlying logic out into the open, and surveying the main kinds of evidence that are used. Such an exercise often proves useful for experts and beginners alike: it never hurts to make sure that one knows what he is doing.

Two Methodologies

For purposes of analysis, we may distinguish two basic approaches to phylogenetical problems. In the first place we have a sort of "formal comparison" or a grouping together of like with like on the basis of structure. The resulting arrangements of organisms need have no historical implications, any more than do classifications of minerals established in the same way. Comparative anatomists long before Darwin were grouping plants and animals into classes on the basis of abstract "body plans" which often enough were thought to reveal the mind of God, rather than the course of history. It is easy to see why such groups turned out to represent, more or less, evolutionary assemblages. The more recent the common ancestry of two organisms, the less the opportunity for evolution to change them. Such static comparison still works; indeed, it remains a part of all phylogenetical research. But contrary to what some authorities suggest (Remane, 1952; Blackwelder, 1964), it is not the whole story. A phylogenetic theory is more than just systematics of the pre-Darwinian kind with an historical explanation superimposed upon it. It involves a second aspect, historical reconstruction. One asks what has actually happened, and finds out. The process of evolution occurs in a reasonably orderly fashion. Understand the laws which govern it, and one has reason to infer that one event is more likely to have occurred than another. The same basic operation is used in all historical sciences, especially geology. In biology, however, it was not applied until the discovery of evolution (Ghiselin, 1969a).

Formal Comparison

In this approach one compares, point by point, the arrangement of parts in complex structures. One tries to line up what appear to be equivalent components in the same basic order: heads, bodies, tails, etc. This gives a basic scheme, or "body plan," for groups and subgroups of organisms. One arranges his materials into classes and subclasses, in such a manner that as one goes from more inclusive groups to less inclusive ones, the body plans become increasingly detailed, with more and more features common to all their members. From among the possible systems, one selects that arrangement which requires the fewest body plans, or (translating into phylogenetic terms) the one which could have been produced by a minimal number of evolutionary steps. Generally it is preferable to base the comparison on structures with a large number of parts, because coincidence and convergent evolution are then less likely to provide misleading similarities. Whole vertebrate or arthropod skeletons are very useful, but entomologists often stress the intricate patterns of veins in insect wings, while new techniques have lately made a number of large protein molecules available. In fact, any feature with a fair degree of complexity is likely to prove rewarding: biochemical pathways, developmental patterns, even behavioral movements. However, it is always wise to bear in mind that we are interested mainly in the order of, and relations between, the parts. We do not merely count similarities. Much as the message in a word is not simply contained in the letters, the historical vestiges in an organism are not embodied in isolated amino acids (Sibley, 1962).

Homology. The terms "homology" and "analogy" are frequently used when distinguishing between the results of common ancestry on the one hand, and those of convergence on the other. The precise meaning of these words is highly controversial, and in fact they are used in various ways. In broad terms, "homology" designates a more fundamental correspondence between parts, such as that which connects the right humeri in vertebrate forelimbs. Analogy refers to a more superficial correspondence, such as that between different kinds of wings: those of birds, bats, pterodactyls and insects resemble one another, but only vaguely, not in details of structure. The main reason for these superficial resemblances is of course evolutionary convergence, caused by the fact that the corresponding parts have the same function. Hence the term "analogy" is generally defined as the relation between parts having the same function (Owen, 1848: 7). I have criticized such definitions because it seems to me that a more basic concept is involved (Ghiselin, 1969b). Superficial resemblances may be caused by influences other than community of function: mimicry is one example. Likewise, parts with the same function do not necessarily share a common morphological pattern. A mushroom's poison and a snail's shell both have the function of repelling predators, but they have little else in common. The most one can say for some analogies, therefore, is that we can somehow relate them to function. Analogues which do have the same function are a special case. This concept of analogy is by no means out of line with common usage, although it is somewhat broader than the usual morphological sense of the term. However, it hardly constitutes a real innovation, since anatomists have been using it ever since Aristotle.

Homology also has broader as well as narrower meanings. It refers to quite a variety of relationships, which are best treated as special kinds of homology, but which nonetheless all refer to correspondences between parts. One example is "serial homology," the relation between equivalent units when these occur as multiples of a single kind in linear order. Thus, in vertebrates, the humerus and femur are said to be "serial homologues." Likewise "homology" is used to designate the equivalent members in the two sets of chromosomes in a diploid cell: the "homologous chromosomes" pair up. For the present discussion, however, the really significant meaning is "evolutionary homology," which is the relationship between equivalent parts in an evolutionary sequence. When we say that the right humeri in two vertebrates are homologous (in the evolutionary sense) we mean that a precursor of this bone existed in their common ancestor, and that a series of corresponding bones connect this precursor with its two contemporary derivatives. We may find it necessary to qualify a statement of homology to provide some important distinctions. Thus, the wings of bats and birds, although they evolved quite independently, are derived from legs in the common ancestor of the two groups. As anterior appendages, they are homologous; as wings, they are analogous.

Some writers (Boyden, 1935, 1943, 1947; Inglis, 1966; Jardine, 1969; Naef, 1919, 1927, 1931; Remane, 1952) have objected to the evolutionary definition of "homology," but their arguments are not compelling. An alleged circularity in the definition turns out to have existed only in the minds of the critics (Ghiselin, 1966a, 1966b; Hull, 1967). The writers insist that one should compare extant organisms to the idealized "body plan" rather than to a hypothetical "common ancestor." Those who maintain the other point of view argue that since history is being discussed, we might as well use an historical terminology. It is revealing that German "idealistic morphologists" (e.g., Naef, Remane) line up against advocates of the synthetic theory and the New Systematics (Haas and Simpson, 1946; Simpson, 1961; Ghiselin, 1969b) on just this issue. A surprisingly large segment of the biological establishment in Germany remains oblivous to the philosophical revolution that began in 1859.

"Homology" is often confused with "similarity" and with "identity." Parts which are homologous do tend to resemble one another. However, when we take distant points in an evolutionary series, such as the anterior limbs in birds and fishes, they have very little in common, and it is sometimes only by the intermediate organisms that we see the connection. Nonetheless parts remain homologous no matter how much they change. It would, however, be most helpful if we had a term, and even better some kind of quantitative index, expressing the notion of the degree of modification. This could be done in terms of the percentage of identical components in common, as has been proposed by Sattler (1966) and by Neurath, Walsh and Winter

(1967). It is misleading, however, to refer to "degree of homology" or to say that homology has been quantified, when one is actually referring to a different relationship.

From homologies to phylogenetic trees. Phylogeneticists still use the formal approach of sorting out their materials so that a minimal number of changes in supposedly homologous parts will account for the observed differences. A good recent example may be found in the work of Fitch and Margoliash (1967; cf. Fitch, 1970) on the protein cytochrome *c*. They obtained "sequence data" giving the order of the amino acids in this molecule for a wide variety of organisms, and programmed a computer to compare all possible arrangements of molecules into phylogenetic trees, and then to select the tree which could have been produced by the least amount of evolutionary change. Their work actually transcended the purely formal approach in that their knowledge of the genetic code allowed them to compensate for the fact that some substitutions are more probable than others. The results were remarkably close to those obtained by classical techniques, which is hardly surprising, since the method was basically the same in spite of the new materials. At about the same time Camin and Sokal (1965) independently developed a similar computer technique for generating phylogenetic trees, and applied it to more conventional morphological data. Yet here again, the fundamental logical operations were nothing new (Ghiselin, 1966b).

We might ask why a minimal number of evolutionary steps is deemed closest to historical reality. The term "parsimony" has been taken from conventional philosophy of science to refer to this approach: the simplest, or most "parsimonious," tree is considered best. Parsimony is a most controversial issue in logic as well as in phylogeny, and one can easily misinterpret it. Camin and Sokal (1965: 323–24) say, "The correctness of our approach depends on the assumption that nature is indeed parsimonious." Rogers, Fleming and Estabrook (1967: 187) assert that "evolutionary parsimony is the assumption that the processes in evolution are efficient." Actually parsimony is a strictly logical concept, and does not assert that the material universe is a simple entity. Neither does it imply that our conceptions of that universe ought to be simple-minded. When one theory, however complicated, can explain the facts, but another cannot do so without having to be supported by additional hypotheses, the former is preferred because it is more "parsimonious," or simple in the logical sense. In other words, a theory which has to be rationalized is more likely to be false than one which stands on its own merits. Yet it is not at once clear how this principle applies to phylogenetic inference. The crucial point is that it involves an implicitly probabilistic manner of reasoning. The aim is to select, from among the various alternative trees, the one which is least likely to be false. If common ancestry is more often the cause of similarities than is convergence, then the fewer the implied convergences the better. Yet since improbable events do happen, it follows that any phylogenetic tree will be only an approximation (see Rogers, Fleming and Estabrook, 1967: 192). The general arrangement of branches will be correct, but their exact location is quite uncertain. Although useful in drawing broad generalizations, phylogenetic trees are rather unreliable when it comes to particular relationships. When we embrace such a logical concept of parsimony, we see that the canon of preferring the "least number of evolutionary steps" may be highly misleading. In parallel evolution, closely related organisms evolve the same features time and again in separate lineages. The probability of parallelism being exceedingly high, there need be nothing out of line with the principle of parsimony when one accepts a theory which implies this sort of multiple origins. The same may be said of a multiple loss.

Historical Reconstruction

The purely formal approach has distinct limitations, and it is highly desirable to use additional techniques. Formal comparison in no way tells us "which way is up" or "what is branch and what is stem" in a phylogenetic tree. Given no more than a system of body plans, we could just as well derive the other chordates from something like a hummingbird as from something like a hemichordate. Opinions to the contrary notwithstanding, we do have other ways of assessing the relative probabilities of phylogenetic hypotheses beyond simply putting like forms together. We may conveniently divide the additional evidence at our command into the *distributional* and the *functional.*

Distributional evidence includes the fossil record, biogeography, and various aspects of comparative ecology. Some workers actually reject evidence of this sort, or attempt to explain it away as something else (Colless, 1969). Indeed, it has been suggested that fossils do nothing more than increase the number of organisms available for morphology of the strictly formal kind (Remane, 1952). This notion that paleontology adds nothing but ill-preserved specimens should not be dismissed lightly, even though it will not bear critical examination. We must grant that it is misleading when one says, without qualification, that the more ancient organisms are preserved in the older rocks. Of course they are, but the imperfection of the fossil record is always a problem. When a group of animals is present in strata of a given age, we know that that group is at least as old as the rocks themselves. But when a group cannot be found, we can never be sure that it was not preserved, or that we have not simply failed to collect it. Negative evidence, for those who want to be safe, is like no evidence at all. This is especially true of the early stages in the evolution of major taxa, for which fossils are notoriously rare. It is always possible that a more primitive form will appear later in the fossil record than its more advanced relatives. However, let us not get carried away. When we actually are fortunate enough to possess a good fossil record, the sequence is clear and events can be dated as well. Furthermore we can tell where such events took place and why. Fossils are preserved along with traces of their conditions of origin, with habitats, competitors, food, and predators. This allows us to place our reconstructed organisms in the historical context of the ecological situation in which they evolved. Such knowledge of the environmental forces acting upon sequences of populations is exceedingly useful in finding out what has happened to them.

Biogeography is often ignored in discussions of phylogenetical methodology, although Hennig (1950) provides a conspicuous exception. This might seem a bit odd, for in the days when the paleontological record was still fragmentary in the extreme, it was the biogeographical evidence that provided the most compelling argument for evolution (Darwin, 1859). Part of the reason may be its unreliability, especially in groups such as rotifers with their resistant stages and ferns with their small spores which can be moved long distances by wind. Also very old relationships have been obscured by migrations and extinctions. Nonetheless, biogeography can tell us a great deal. The distribution of organisms is determined by their ability to disperse, both by their intrinsic ability to move about and by their opportunities to do so. Slow-moving animals, such as terrestrial snails, tend to have restricted ranges. Those which cannot swim or fly tend to be excluded from remote islands, and their limits of distribution coincide with the obvious barriers. Relatively isolated parts of the world, such as South America and Australia, have evolved their own endemic faunas. Groups which are thought upon other grounds to be very old, yet have limited capacity for dispersal, nonetheless have wide distribution of their higher taxa (e.g., onychophorans). And on a statistical basis, the closest relatives tend not to be far removed from one another in space. To be sure, such rules are unreliable enough that they should be used with particular discretion. But they are still valid, and many systematists use them routinely, especially for taxa of lower rank, such as genera and families.

As a final example of distributional evidence, let us consider the relationships between parasites and their hosts. By extension the same principles can be applied to other ecological data, but it is with parasites that the technique has found its most fruitful application (Dogiel, 1966). Parasites tend to develop a specificity for a narrow range of hosts. Since it is not easy for them to switch from one host species to another, they evolve rather like organs of their hosts' bodies, and each group of animals and plants has a characteristic parasite biota. Hence the phylogenetic trees of hosts and parasites tend to be somewhat congruent. However, certain qualifications are necessary. Parasites do evolve new host relationships. Very often they change to a closely related species, because the adjustment is less difficult; in this case one might get the false impression that the related hosts had derived the parasite from their common ancestor. Furthermore, organisms which live together and have similar habits often develop convergent faunas of symbionts. Man, for example, shares a disproportionately large number of parasites with the rat and the pig. The parasitological approach is most reliable when combined with biogeographical evidence, and when convergence is taken into consideration.

An interesting, albeit still controversial, example of the parasitological method may be taken from the efforts to determine whether the "ratites" (such flightless birds as ostriches) have had a monophyletic or a polyphyletic origin (see de Beer, 1956). Biogeography and vestigial organs show that they have lost the power of flight in more than one lineage, but this does not necessarily mean that they lack a common ancestor which might be considered a flying ratite. Many old-fashioned, typological anatomists viewed their similarities as convergent. Subsequently Bock (1963), using the new functional anatomy, including a novel way of dealing with parallelism, provided evidence that they are monophyletic. The same conclusions were reached independently by Meise (1963), using behavior, and by Sibley (1960) with new molecular data. Half a century earlier, Kellogg (1913) reached a similar position through a study of their ectoparasites. Although his data were rather sparse, and although his views were long contested, his approach seems to have been vindicated.

All these distributional techniques involve a perfectly straightforward principle of inference. They are neither more nor less than versions of the conventional "scientific method." One hypothesizes about what has occurred, and tests the hypotheses by seeing if one can refute them. Space and time are such that certain conceivable relationships are physically impossible, or at least improbable. A population cannot evolve from one which does not yet exist. A parasite does not flourish a thousand miles from its host. By excluding false hypotheses, we arrive at our reconstructed history.

Functional evidence. "Morphology," strictly defined, has nothing to do with function, which is the province of physiology. It deals only with "form" and some biologists try to maintain a purely morphological approach to phylogeny. Comparative anatomists, however, may employ techniques of historical reconstruction in which function plays a decisive role. The approach is basically the same as that used with distributional evidence. One asks if hypothesized events are consistent with biological laws and principles. If not, they are rejected as false. An inviable common ancestor can only mean that a different one must be reconstructed. If selection or another evolutionary mechanism is incapable of producing an hypothesized change, we prefer some other hypothesis. Note, however, that we reject the hypothesis because it is contradicted by some law of nature. We do not accept it merely because it is consistent with such laws. The reason is a simple principle of logic which we all use in our daily lives. A jury does not convict a defendant solely because he could have perpetrated some crime, but it will always acquit him if he could not possibly have been guilty. Similarly, all correct hypotheses about phylogeny must be consistent with the laws of nature, but to say that one hypothesis conforms to such laws tells us nothing about alternatives which meet the same criterion. Thus some of Hanson's (1963) arguments for the ciliate origin of the Metazoa are invalid, for all they establish is that multicellularity has advantages; but a multicellular "flagellate" would derive equal benefit. An experimentalist might say that the null hypothesis was not contradicted.

One of the more valuable kinds of functional reasoning involves the use of "vestigial organs." These are parts which are not used at present, but which are interpreted as leftovers from an earlier stage in which they were functional. The internal tail of human beings, which retains the muscles necessary to wag it, is a good example. When such structures are present we have traces of earlier conditions not unlike the fossil record—with the historical vestiges preserved in DNA rather than in rock. We also have a way of telling "which way is up," something which (as was noted earlier) is lacking in a purely formal approach. It is easy to see how our ancestors' tails could have gradually been reduced through natural selection, once they had ceased to be useful. However, a change in the opposite direction would require the gradual enlargement of an organ through long periods in which it was adaptively worthless. Since there can be no selection for a benefit realized only in the distant future, no such changes can occur. A phylogeny which derived tail-bearing primates from ancestors with a condition like our own would be a poor hypothesis indeed.

There are some difficulties with vestigial structures, particularly with how one tells that they lack a function. Actually at least some may still play one role or another in morphogenesis. Nonetheless, certain features, such as the cutting edges on vestigial teeth, are not used in the ordinary way. It has been denied that one can demonstrate lack of function in a part (Rudwick, 1964b), but criteria for doing so are in fact available. Darwin (1859:149) points out that vestigial structures are highly variable. Wallace (1889:217) noted a loss of symmetry in the color patterns of domestic animals. Remane (1952) refers to a kind of disorganization. All this makes sense from a theoretical point of view, for stabilizing selection ought to keep parts that are still used within a narrow range of size and form.

Much functional reasoning tends to be somewhat less rigorous. Hypotheses are rejected, not because they are impossible, but merely because they seem improbable, or because one seems more plausible than another. Often relative probabilities are exceedingly hard to estimate, and one easily slips into very subjective habits of thought. Nonetheless, certain "rules of thumb" have proved reasonably trustworthy in practice. Many of these have to do with steering clear of situations in which convergence is likely, while others help to determine the direction of evolutionary change. Among the most useful is the principle already mentioned that a complicated structure is more easily lost than gained (Brown, 1965). The evolution of a new organ often requires a number of unusual coincidences. On the other hand, the loss of a part is just about inevitable when it ceases to be used. Wings, for example, have evolved only a few times in the history of life, but they have been lost time and again.

Convergence is especially probable wherever different lineages all come under the influence of the same group of selection pressures. The resemblances between fish, ichthyosaurs and whales are notorious. If one knows how organisms function under particular circumstances, it is much easier to avoid being misled by such convergences. One looks for features of the body which evolve independently of the parts subjected to the common selection pressures. Thus, where only locomotion is affected, one infers the relationships on the basis of such structures as the reproductive system.

On occasion it has been maintained that convergence is no serious problem, because we have instances of its being detected (Sokal and Sneath, 1963:106–107). Others claim that since convergence is always possible, it poses an insuperable obstacle (Pantin, 1966). The answer to such extreme positions is twofold. On the one hand, where one *can* demonstrate that convergence has taken place, the issue of separate origin is thereby settled. But given identical structure, the possibility of convergence always remains with us, and one cannot infer that it has not occurred. The problem is most pressing in the case of parallelism, that is to say, where several closely related lineages evolve the same adaptation from the same original condition. If interconnecting forms are not available, then one tends to assume that the adaptation has arisen only once. On the other hand phylogenetic inferences need not be based on convergent structures, nor on those which evolve in a single direction. The most reliable phylogenetic trees are derived from structures which have diverged, and in which there has been an adaptive radiation. Parallelism generally occurs where the organisms have only one way of adapting to a given selective influence. When a variety of responses to an environmental challenge are possible, divergence is more general. Parallel trends in one kind of feature may occur side by side with divergent ones, and we need only weight our evidence in favor of the divergences to obtain a good picture of the relationships. Thus in my own work on opisthobranch gastropods (Ghiselin, 1966c) I had to deal with a group in which the shell has been lost polyphyletically, so that snails have repeatedly evolved into slugs. On the other hand, a clear-cut adaptive radiation may be detected in the feeding habits, the genitalia evolve in different directions, and the proportions of the body in divergent lineages are not the same. (See Bock, 1969, for additional examples.)

Another important functional technique is the search for likely pre-adaptations. Really new structures are very hard to evolve. Often a part originates by the modification of some pre-existing one, with a new function being either added or substituted (*Funktionswechsel* in German, there being no English equivalent). In order for this to happen, the precursor must have such properties that it can be pressed into service for the new function. Much of the voluminous literature on the coelom (see Clark, 1964) focuses upon just this point. The coelom, or secondary body cavity, may have been evolved from any or all of several sources. Most theories as to where it came from have one feature in common: they try to derive it from some other cavity. Thus, different workers have traced it to the lumen of the gut (enterocoel theory), the kidney (nephrocoel theory) or the gonad (gonocoel theory). The exception is the schizocoel theory, which, because it considers the coelom to have arisen *de novo*, is for this very reason less plausible, although for other reasons it is widely advocated. The technique here is obviously a case of historical reconstruction. It also illustrates another important point, that much of the discussion about the relative merits of "adaptive" and "non-adaptive characters" is irrelevant. We are asking how organisms have evolved, not just grouping them together on the basis of shared attributes.

Various cautionary measures are wisely adopted when using a functional approach. Various supposed laws of nature have been either intemperately applied or uncritically rejected. The "biogenetic law" of Haeckel ("ontogeny recapitulates phylogeny") has been more than adequately debunked (de Beer, 1958). Embryological development is not a simple repetition of evolutionary history. Yet embryology is still useful. Modern studies increasingly emphasize the dynamic aspects of development. If evolution occurs through modifications in developmental processes, as indeed seems to be the case, then understanding morphogenetic mechanisms should help us to determine which historical changes are most likely to have occurred. However, an excess of confidence and inattention to methodological difficulties detract severely from much embryological reasoning. The use of negative evidence from embryology is one of the most frequent vices of phylogeneticists. When a vestige of some organ appears in development, we know that an ancestral form possessed it. But when it is lacking, we cannot decide whether or not it has ever been present unless we have additional evidence. Chicken embryos have no teeth, but the fossil record of early birds shows that their ancestors had them.

Dollo's law, which states that a complex feature cannot return to an earlier state, is true, but only in a qualified sense (Gould, 1970c). Of course, when whales and ichthyosaurs became aquatic, they came to resemble fish, but they retained many features of their terrestrial forebears. Even less reliable has been the application of Cope's rule, which states that evolution proceeds in the direction of increasing body size. Although certain lineages do manifest trends in the production of increasingly larger organisms, even these show many exceptions, and major lineages in some groups go the other way (e.g., rotifers and copepods). The whole issue of how one deals with scale effects is most complicated, for increased and decreased size both produce a host of convergences. Thus bodily enlargement gives rise to such features as graviportal limbs, shifts from ciliary to muscular movement, and the folding of secretory, absorptive and other surfaces. Reduction leads to numerous simplifications and to a larger proportion of the body being devoted to such crucial functions as the activities of the nervous and reproductive systems. Many older taxonomic groups have turned out to be little more than size categories. It is a good sign of our sophistication that such mistakes have become increasingly less common.

A final problem with the functional approach is a sort of teleology, in which one assumes that evolution always does the right thing and that all characters are adaptive. Of course in a sense we can view whatever evolutionary changes do occur as adaptive in one way or another. Nonetheless populations lack foresight, and they aren't always able to do what seems appropriate to us. Ax (1966) has drawn attention to the bizarre complex reproductive systems of certain flatworms, many features of which seem hard to explain as the results of adaptation. Likewise Darwin (1877) used what we might call "Rube Goldberg devices" which orchids have evolved as pollen transfer mechanisms, to show the absurdity of naive teleology in general. In my own work on gastropod reproductive systems (Ghiselin, 1966c) I experimented extensively with various functional techniques. I started out with the working hypothesis that the organisms would show "functional improvement." But experience soon taught me the limitations of my methodology. To be sure, I could find plenty of adaptive trends, such as grooves turning into ducts and separate channels coming to handle eggs and sperm. But one group of slugs showed a series of changes that hardly made any physiological "sense" at all. And some appeared, so to speak, to have "given up," and had shifted to copulation by hypodermic impregnation, rather than accomplish it via the usual orifice. Perhaps the functional reasons for some of the changes in complex reproductive systems would become apparent with more study or a better imagination. One possibility is that shifting the positions and connections of ducts has provided isolating mechanisms. An alternative seems to me equally plausible: pleiotropy. That is to say, a gene with more than one phenotypic effect might be selected for just one of these, and as a result the other effects would evolve as adaptively neutral or even deleterious side effects. I have some reason to suspect that pleiotropy is indeed one of the important influences, but the evidence is far from decisive. In some of my slugs, the prostate and another secretory organ become elongated and tubular rather than compact, for no obvious reason.

But this change happens in a group in which the digestive gland has undergone the same modification. Can it be that genes for producing tubular structure have done more than is really necessary?

Such experience reinforces my conviction that we need a multidisciplinary approach to paleobiological problems. Paleontology and neontology have different strengths and weaknesses, owing to the nature of their materials. Fossils are dead, good ones are rare, and they derive from only part of the original animal or plant. Function is much harder to infer in fossils than in living specimens, because we can't watch the organs at work. A paleontologist is lucky when he has enough of a series to deal with population dynamics. And the organ systems he deals with are mainly hard parts, such as teeth and skeletons. His materials predispose him to view evolution from the perspective of survival, especially predator avoidance and feeding. Neontologists, however, have much better access to reproductive structure and function. Hence they find it much easier to deal with evolution in terms of isolating mechanisms and differential reproduction. One result has been the disagreement between the two groups of scientists over species concepts. Another may be even more important, because it is not so obvious. The reproductive apparatus, which rarely fossilizes, seems to evolve in a manner somewhat different from that of other organ systems. Its adaptations are less immediately useful, and at least some maladaptation occurs. Our continuing synthesis will require the cooperation of persons with a diversity of outlooks and resources.

Functional approaches to the interpretation of fossils. In strictly formalistic works, diagrams called "archetypes" are often presented. These are not, as one might think, "models" of common ancestors. Rather they are efforts to summarize, in diagrammatic form, the features common to all members of a taxonomic group. They are "pictorial diagnoses," not representations of organisms. In historical approaches to phylogeny, however, similar constructs are intended to depict, as closely as possible, real animals and plants. They are models in the sense of being hypotheses about concrete organisms, accepted or rejected according to how well they fit the scientific evidence. As with all hypotheses in the historical sciences, we have uncertainties that arise from the necessity of extrapolating backward in time from contemporary materials. Paleontologists are often in a better position than are neontologists, because certain of these extrapolations are not necessary. A myth has therefore arisen that the paleontologist can, or should, read the history of life directly from what he perceives in the fossils, without theorizing or speculating.

Yet in fact paleontologists infer a great deal more about their materials than the brute facts would seem to justify. They often go far beyond the superficial appearances to reconstruct the missing soft parts, the ecology, and even the behavior of fossil organisms. In so doing they use much the same kind of reasoning as do neontologists, who likewise have to think before they know what an organ does. There is no question that some applications of functional analysis have been amply rewarded. We know a great deal, for example, about the musculature and functional anatomy of fossil vertebrates, thanks to the insertions on the bones. Likewise teeth allow one to infer a great deal about diet. When we study the functional anatomy of contemporary organisms, we can find out what their parts do, and how these work, by observing them in action and by experimenting. With fossils we must rely on somewhat more "indirect" criteria, but nonetheless the principles are the same. A tooth is easily recognized by the fact that its structure is appropriate to biting or crushing, by its position, or by its connection with the jaw. On the other hand, the interpretation of fossils is often hampered by inadequate understanding of functional anatomy. Oversimplified ideas about the role of notches in the shells of mollusks have led to much confusion about the internal structure of gastropods and their precursors (see Rollins and Batten, 1968).

Another problem arises with incorrect applications of the principle of uniformitarianism (see Kitts, 1963; Gould, 1965). In its most general formulation, uniformitarianism demands that historical events should be reconstructed by the application of laws that are universal, and hence apply irrespective of time and place. In part it seeks to get away from *ad hoc* rationalizations; it also gives rigor through grounding inferences upon reliable premises. More loosely formulated, it suggests that "the present is the key to the past." So far as this means only that we should derive some of our premises by studying what goes on about us here and now, all well and good. It is much easier to investigate many phenomena when we use contemporary materials. If, on the other hand, it is meant that generalizations which apply for organisms here and now must necessarily apply to all other organisms, then clearly we go too far. If animals and plants have in fact evolved, some of their present features must differ from those which existed in the past. For instance, the pulmonate gastropods include the familiar, air-breathing snails and slugs of land and fresh water. Had the ancestral stock died out, it would have been much harder to realize that the group arose in the sea and invaded the land polyphyletically. Indeed, early workers reasoned from what they knew of non-marine forms, and wrongly concluded that the aquatic ones had once been terrestrial. This notion is still perpetuated in the literature, even though the evidence against it (see Marcus and Marcus, 1960) is overwhelming. The diagnoses of systematics are not laws of nature. They are statements about the consequences of historical events. Hence there is no reason to think that one animal will inevitably have the features that are present in the other members of its group.

Conclusions

The study of phylogeny, whether it uses fossils or not, is a strictly objective branch of natural science. Hypotheses are proposed, developed, and tested, much as they are in physics and chemistry. The methods appeal to experience—to the rocks and to the organisms—not to opinion or to authority. Occasional pronouncements to the contrary show only that the scientific method has been inadequately appreciated. If lack of data leaves many uncertainties, the same may be said for any growing branch of knowledge. In the next few years, we may be confident, the accumulation of sequence maps for proteins will resolve what remain of our difficulties with the relationships between phyla. The anatomical work of the last hundred years will have served its function, in giving a first approximation and a guide; its basic findings in all probability will not prove grossly in error. Anatomists, knowing what the phylogenetic relationships actually are, will then gladly shift to a new series of problems, such as explaining convergence rather than detecting it. Paleontology will retain its value in providing dated materials for systematists, but its phylogenetic interests too will become less descriptive and more explanatory.

Like stratigraphy, phylogenetics is an historical science. It helps to bridge the gap between geology and such disciplines as chemistry which deemphasize time. We may hope that as knowledge advances, the importance of the past will become more generally known. Now that molecular biology has solved many of its original problems, ones that were concerned with mechanisms, molecular phylogeny has become increasingly fashionable. An awareness of history will perhaps gradually force itself upon chemistry and physics, and upon the philosophy of science, which has all but ignored it. Phylogeny in itself provides a fascinating exercise in inductive reasoning, but it has more than just intrinsic interest. Fundamental to the evolutionary world-view, it permeates our thinking about organisms. Its central position assures us that whatever the future vogue of the subject, phylogeny will continue to challenge our ingenuity.

PART IV

Distribution

8

CONCEPTUAL MODELS OF BENTHIC MARINE COMMUNITIES

Ralph Gordon Johnson

Editorial introduction. The interpretation of recurring associations of animal and plant fossils as biological communities—and hence significant of something more than accidental occurrences—was put on firm ground for invertebrate fossils by Max Elias in 1937 in a paper far ahead of its time. Curiously, this approach was not widely used until A. M. Ziegler's seminal paper of 1965 on Silurian associations signaled a major line of research in the 1960's. During this period, many paleontologists have used this approach to bring order out of chaos in faunal collections of many geologic periods.

The analysis of recurring associations in the fossil record now has proceeded to other questions. Ralph Gordon Johnson proposes a way in which sequences of changing communities may reflect sequences of semi-disasters, and in this sense each separate community represents part of a continually changing pattern of ecologic succession. For Johnson, the immediate history of a fauna is critical. Within the general theme of succession, he develops a model that emphasizes precision and reality in its predictions. This approach promises to open the way toward precise understanding of spatial and temporal variation in paleontologic communities of marine invertebrates.

Introduction

The interpretation of recurrent fossil assemblages as fossil communities has become a common practice in paleontology. The use of these interpretations in stratigraphic or evolutionary studies is dependent upon our conceptual models concerning the organization of the kinds of communities involved. For example, if we believe the species of the community are highly mutually interdependent, then we are likely to emphasize biological interactions in interpreting faunal changes. On the other hand, if we believe the species are relatively independent, then we might emphasize the biological consequences of changes in the physical environment. Thus, our view of the nature of the community under consideration will determine the kinds of observations we might wish to make. If we conceive of the community as biologically integrated, we might analyze the fauna in terms of food relationships. Or if we regard the community as being largely controlled by the physical environment, then we might seek lithologic and stratigraphic evidence of environmental change.

The paleoecologist is interested in explaining the occurrence of particular fossil assemblages. For any assemblage we might ask: "Why are these species here, at this place and in such numbers?" We are further interested in how and why the fauna appears to change laterally and stratigraphically.

Modes of Explanation

Lewontin (1969b) has suggested that there are often two opposing modes of explanation in population biology. There is an equilibrium mode describing the equilibrium properties of the ecosystem without recourse to historical information. A good example is the mathematical expressions relating such general properties of the community as diversity, biomass, productivity, and stability (Margalef, 1968). Such models are extremely important to our understanding of the general dynamics of the community.

On the other hand, one can recognize an historical mode of explanation, an explanation of the state of a population or a community by reference to specific historical events. Such models can deal with particular communities and are important because they imply the possibility of predicting future states in terms of particular species. Obviously, this kind of explanation appeals to paleontologists.

An historical explanation of the state of a community would involve both organic and inorganic evolutionary factors. On the grand scale, the occurrence of a particular suite of species is determined by a history of organic evolution and an inorganic history resulting in a unique geologic, geographic, and climatic configuration.

On a fine scale, we can also recognize non-evolutionary (or contemporary) factors such as larval dispersal, substrate relations, predation, competitive exclusion, etc. The distinction between evolutionary and non-evolutionary factors is really arbitrary since, to the degree to which contemporary factors influence gene frequency, they are also evolutionary.

Usually it is unnecessary to take into account evolutionary factors to explain the spatial and temporal variations observed in the modern environment. We can neglect such factors in predicting faunal changes following an environmental disturbance. With care, we can neglect evolutionary factors to explain faunal changes in local stratigraphic sequences.

In recent years there has been a considerable interest in the relations between species diversity and environmental stability (e.g., Slobodkin and Sanders, 1969). I believe that these relations can serve as a basis for a conceptual model of the community that can be used as an historical explanation of non-evolutionary faunal changes at a particular place. In this paper I will attempt to develop such a model for benthic marine communities.*

Nature of Shallow Water Benthic Marine Communities

There has been a long dispute as to whether communities are discrete, discontinuous units or continuous gradations of one species assemblage into another (Gleason, 1939; Cantlon, 1968; Mills, 1969). In a recent study of Tomales Bay, California, I found that the benthic species could be readily sorted into three intergrading communities corresponding to the general nature of the clastic substrate (Johnson, 1971). The distributions and associations of the species are a continuum without sharp transitions or natural breaks. This circumstance has been reported for other benthic communities in clastic substrates (e.g., Lindroth, 1935; Lie, 1968). On the other hand, most benthic communities on hard substrates and many terrestrial communities appear to have sharp boundaries.

It seems to me, and others, that from such observations and theoretical considerations, the answer to the question of whether communities are discrete or continuous is yes! Some communities are continuous and some discontinuous depending on the slope of environmental gradients and on the complexity of the community. There is a complete spectrum in nature.

---

* Since the benthos is a part of the ecosystem which includes the overlying water mass, the term "benthic community" could be regarded as a misnomer. However, it is now a conventional term. No ambiguity results as long as it is realized that the distinction is arbitrary. As stated by Margalef (1968, p. 13): "Any ecosystem under study has to be delimited by arbitrary decision, but one has to remember always that the imposed boundaries are open. . . ."

If the environment is gradational, the communities will tend to be gradational as well (Beals, 1969). If there are sharp breaks in environmental gradients, the communities will appear to be discontinuous. The environment of benthic communities on clastic substrate is usually gradational. Muds grade into muddy sands and these, in turn, into sands. Depth, light, salinity, and water motion tend to vary in a continuous fashion. As a consequence, benthic communities on clastic substrates tend to be continuous and intergrading, both in shallow and deep water (Sanders and Hessler, 1969). Hard pavement substrates are often more sharply limited, ending abruptly in clastics with a sharp change in fauna.

Another factor is the ecological complexity of the community. Margalef (1968) has suggested that the changes that occur during ecological succession can be used to identify the stage of maturity of communities existing at a particular time and place. According to Margalef, low grade or immature communities are characterized by low species diversity and fluctuate in composition under the direct impact of the physical environment. A mature or high grade community can be recognized by high diversity and a relative constancy in composition. These community grades are similar to Sanders' concepts of physically controlled or biologically accommodated communities (Sanders, 1968).

If low grade, physically controlled communities contain relatively few interdependent species systems, they would recur in variable combinations of species depending upon the vagaries of recruitment and environmental heterogeneity. Such communities might grade continuously into adjacent low grade communities. High grade communities probably contain many interdependent species and therefore would tend to recur as discrete units. We would expect such communities to have fairly abrupt boundaries with little intergradation into proximal communities.

Little is known about the interdependence of species living in clastic substrates. Except in a few instances, such as the famous case of *Urechis* and its commensals (Ricketts and Calvin, 1952), most species tend to recur in variable combinations. In pair by pair analyses, it is often possible to assemble groups of significantly associated or correlated species (species A is significantly associated with species B, and both with C, etc.). However, the groups of species defined in this manner may not actually occur together very often in nature. For example, Juskevice (1969) made a study of interspecific correlation and association in the benthic marine communities of Tomales Bay, California. One group of 11 highly intercorrelated species occurred together as a unit at 13 of 177 stations. Another group of 4 species was found to occur together only once. Similar results were obtained using groups assembled from species associations defined by a Chi-square test of independence.

In these regards, benthic communities on clastic substrates in shallow water are relatively low grade communities. They exhibit low diversity, recur in variable combinations of species, and are often revised by fluctuations in the physical environment. The environmental gradients tend to be low. As a consequence, benthic communities on clastic substrates tend to be contînous and intergrading.

From this viewpoint, a benthic community is most conveniently defined as the assemblage of species at a particular place and time. As environmental factors recur, there will be a tendency toward the recurrence of particular species but in variable combinations.

Paleontological Implications

Horizontal and vertical gradations of fossil assemblages are commonly observed in continuous stratigraphic sequences. Abrupt faunal change is usually explained as due to abrupt changes in the environment of deposition. On the other hand, there has always been a tendency to explain intergradation as the result of *post mortem* transportation. *Post mortem* transportation has probably been unduly emphasized. More and more studies have shown that transportation is rare even in relatively high energy environments today (Straaten, 1960; Valentine, 1961; Johnson, 1965; Warme, 1969). When appreciable *post mortem* transportation does occur, the skeletal remains are often associated with sedimentary structures no self-respecting geologist could overlook. It seems to me that most of the intergradation of fossil assemblages reflects the intergradation of communities. It is time that paleontologists stop apologizing about this possible source of bias as there are enough probable sources to worry about.

Faunal changes in a constant environment. We do observe changes in species composition in modern communities without any noticeable change in the overall physical environment. Similarly, we commonly observe changes in continuous stratigraphic sequences that are not accompanied by any obvious change in lithology. Some of these may be the result of the vagaries of recruitment. More often such faunal changes probably are the result of the ecological interactions involved in succession. If we could know the local order of succession for a particular ecosystem, we could predict this kind of change. One possible conceptual model suggests that the information we need is always present in the system.

Consider the following propositions:

(1) Any disturbance that affects the species composition will downgrade the community to an earlier stage of succession.

(2) Distant parts of a benthic community are independent of one another since the interactions of most species occur over very short distances and during short periods of time.

(3) A local disturbance will downgrade part of the community without affecting other parts.

(4) Small-scale disturbances are continually occurring in natural populations.

(5) The community is therefore a temporal mosaic, parts of which are at different levels of succession.

(6) Spatial and temporal variations in species composition will be large in communities occupying harsh or unpredictable environments.*

(7) Spatial and temporal variations in species composition will be small in communities occupying benign, predictable environments.

These theoretical propositions are largely derived from generalizations concerning ecological succession (Margalef, 1968), the relationship between species diversity and environmental predictability (Slobodkin and Sanders, 1969), and the view of the nature of benthic marine communities described earlier. They can be used to reconstruct the local order of species in succession.

Since variations in space are similar to temporal variations (the community is a temporal mosaic), the order of succession can be approximated from a study of spatial variations. As low diversity is one characteristic of earlier stages of succession, we can simulate the order of succession by placing samples taken from the community in order from low to high diversity. We will obtain the greatest range of diversity with the fewest samples if we sample along a stress gradient. The lowest diversity will be observed in the more unpredictable parts of the environment while the highest diversity will be observed in the more predictable regions occupied by the community.

I have used this theory to explain temporal variations in species composition of an intertidal community over a five year period (Johnson, 1970). Recently, as a result of a change in the course of a tidal channel, a large wave of sand moved slowly across the lower part of the intertidal study area. The species assemblages low on the tidal flat were downgraded in the rank order predicted by the previous study. In this instance, the local disturbance was the rapid deposition of sand. Deposition and erosion are probably very common sources of the general kind of disturbance postulated in the model.

The concept of the community as a temporal mosaic can be diagrammatically represented as in *figures 8–1* and *8–2*. This concept is at variance with the commonplace notion of the balance of nature. Far from attaining some equilibrium state, the community is conceived of as continually varying in response to a history of disturbance. In this view, the community is a collection of the relics of former disasters.

---

* The term "unpredictable environment" is used here in the sense meant by Slobodkin and Sanders (1969). An unpredictable environment is one "in which the variances of environmental properties around their mean values are relatively high and unpredictable both spatially and temporally" (p. 83).

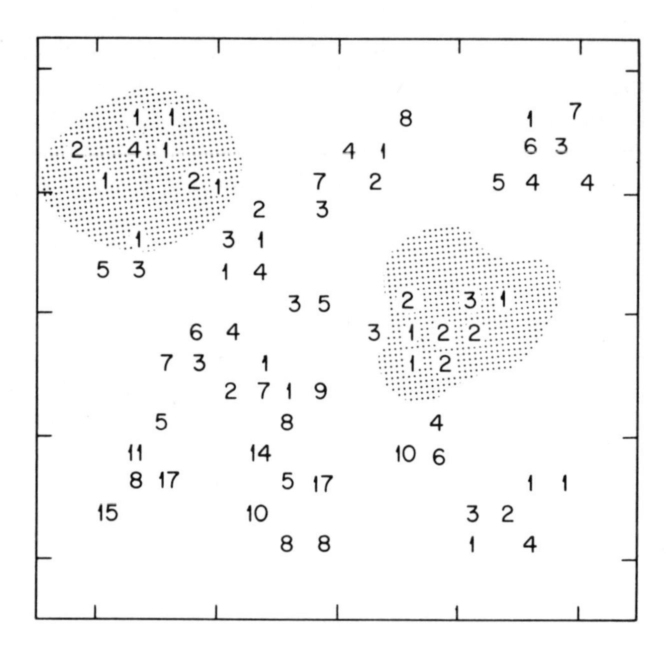

Figure 8–1: Hypothetical map of a benthic marine community in a homogeneous sediment. Individuals are shown as numerals. The value of the numeral indicates the rank of the species in the order of succession. The dotted areas are those most recently disturbed.

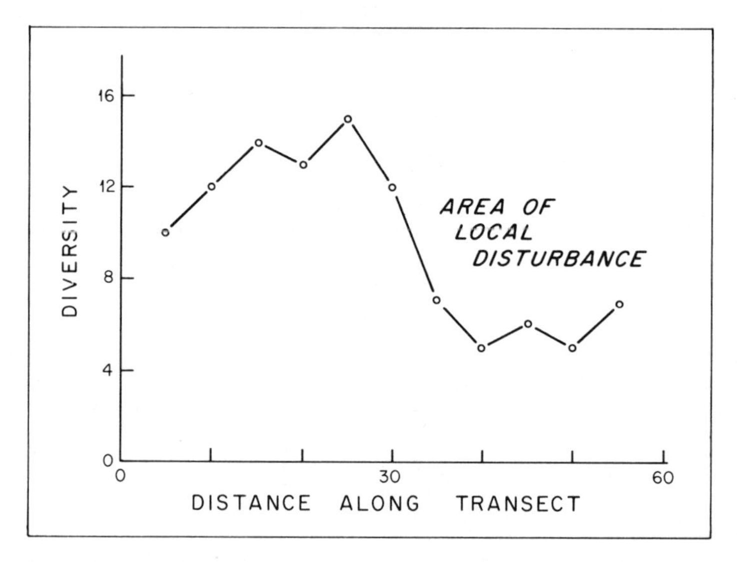

Figure 8–2: Hypothetical transect through a benthic marine community in a homogeneous sediment. The units of length and species diversity are arbitrary.

There may be some objection to the use of the term "succession" in this context. As usually defined, succession is the replacement of one community by another. However, this process must proceed species by species in a continuum of change. I can see no compelling reason to distinguish between a "microsuccession" and a "macrosuccession" under these circumstances.

The common explanation of succession is that the community modifies its environment in such a way that eventually it becomes less suitable for the existing community. Its members can no longer successfully compete with invading species of the next stage. In this way, earlier stages prepare the way for later ones. The details of this process have not been worked out for benthic marine communities. There can be no doubt, however, that benthic communities do progressively modify their physical environment.

A newly deposited sediment is a relatively homogeneous habitat. As organic detritus accumulates at the sediment-water interface, the animals rework it into the underlying sediment. Through their feeding and burrowing activities, the sediment becomes increasingly heterogeneous on a small scale. The particle size, composition, and stability of the sediment may be altered (Rhoads and Young, 1970; Rhoads, 1970). In addition, the activities of micro-organisms must profoundly affect the chemical environment. The accumulation of shells and burrows would further add to the heterogeneity of the infaunal environment. Over a period of time, however, this heterogeneity may be reduced by the more thorough reworking of the sediment (Moore and Scruton, 1957).

These modifications of the sediment must be accompanied by faunal changes. We would expect that the attractiveness of the substrate to settling larvae would change. We could also expect that the fine scale heterogeneity developed by the activities of members of the community would affect species diversity in the following way. In the beginning, species diversity would be low in the relatively homogeneous sediment. As the sediment becomes progressively more heterogeneous, providing more microhabitats, diversity should rise. Species having slightly different habitat preferences could live side by side. Mangum (1964) found five maldanid polychaetes, differing in habitat preference, living together in a conspicuously heterogeneous sediment. When the sediment becomes more thoroughly mixed, we would expect diversity to drop off to a somewhat lower level. If a large quantity of sediment is added rapidly to the area, the community would be downgraded to an earlier stage of succession. However, the addition of small amounts at low rates might maintain a high level of heterogeneity while at the same time adding food materials. From this viewpoint, sedimentation is a necessary condition for the maintenance of diverse benthic marine communities.

The conceptual model described here concerns the variations in species composition within a benthic community in a relatively constant general environment. Similar concepts can be used to explain the changes in species composition as the environment changes. Adjacent communities are most likely those which will replace one another as the environment changes. Since adjacent benthic marine communities tend to intergrade in space, the relationships between coexisting communities might be used to predict how they intergrade in time. This concept is identical to Walther's Law in stratigraphy. Walther's Law points out that lateral facies changes are commonly seen in vertical succession as environments of deposition move back and forth in time.

Faunal changes in a changing environment. In a recent investigation of animal-sediment relations in Tomales Bay, California, I found that most of the species studied had a clearly preferred substrate (Johnson, 1971). However, some individuals of such species were occasionally found elsewhere. There was a striking tendency for species occurring outside of their characteristic environment to be associated with the most diverse assemblages of a foreign substrate. Furthermore, after reconstruction of the order of succession on a particular substrate, as described here, it was found that the species low in the order of succession were those species that are found most frequently on other kinds of substrates. These circumstances can be represented diagrammatically as in *figure 8–3*.

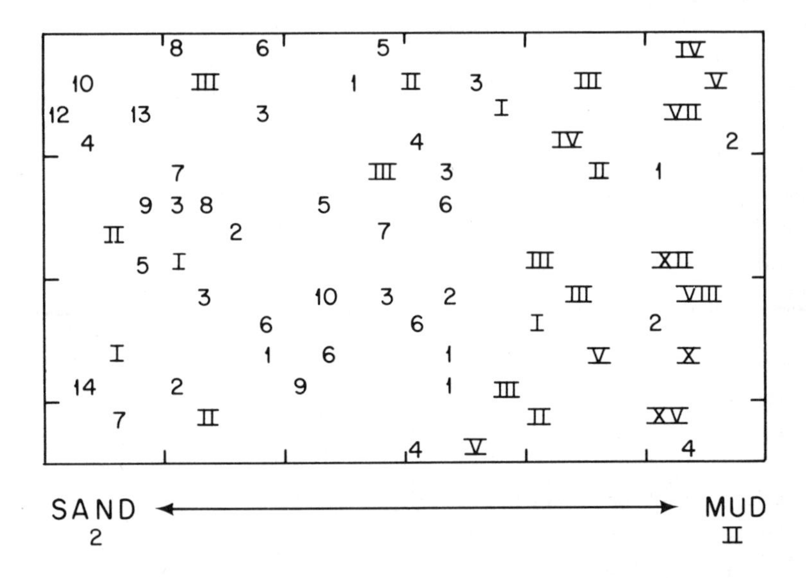

Figure 8–3: Hypothetical map showing the distribution of species in the intergradation of mud and sand communities. Individuals are shown as numerals. The value of the numeral indicates the rank order of the species in succession. Mud species are represented by Roman numerals and sand species by Arabic numerals.

We would expect that the species appearing early in a successional sequence would be eurytopic. Their association outside of their characteristic substrate with diverse assemblages can be at least partially explained in terms of environmental stability. If a larva manages to settle and metamorphose in a foreign substrate, it probably has a greater chance of survival in the more predictable (hence supporting more species) parts of the environment in spite of specialized predators and strong competition. The concept of the community as a temporal mosaic suggests that most sites are not at equilibrium and are undersaturated with respect to the number of species that could occupy them.

In the Tomales Bay study, two polychaete species were found to be truly ubiquitous in the sense that they occurred in every major subtidal environment at about the same level of abundance. They did not have a characteristic association with either a type of substrate or an assemblage of species. These two ubiquitous species were also significantly associated with the more diverse assemblages throughout the bay.

From these observations and the previous considerations, it might be useful to consider a benthic community on clastic substrates as containing three kinds of species:

(1) characteristic species—those that occur more often and in the greatest number in the particular environment, and

(2) intergrading species—those that are characteristic of another, usually adjacent environment, and

(3) ubiquitous species—those which occur in several environments but are not characteristic of any one.

These distinctions are conceptual and not meant to imply that all three kinds of species are not members of the community. Obviously they are part of whatever community they are living in.

It may be the general case that intergrading species tend to be low in the order of succession in their characteristic environment. In addition, intergrading species tend to occur in the most predictable (i.e. diverse) parts of the community. If so, then these proposals could be used to predict faunal changes in a changing environment. The whole conceptual model can be described as follows:

(1) Under normal circumstances, the species composition at a particular place will fluctuate in response to local environmental fluctuations (disturbances).

(2) The order of appearance and disappearance of species will tend to follow the local order of succession or regression.

(3) As the environment changes, the community will be downgraded to lower order, characteristic species.

(4) Intergrading species from proximal communities will appear more frequently among the most diverse assemblages of the community.

(5) As the community continues to be downgraded, these intergrading species will spread throughout the region undergoing change.

(6) As the environment continues to change, the species distinctive of higher stages of succession in the new environment will make their appearance and spread.

Applications of the Model

The propositions described here can be tested by field experiments or by monitoring natural and artificial disturbances. They appear to be consistent with generalizations concerning succession and the relations between species diversity and environmental predictability. There is some evidence in support of the propositions concerning spatial variations within benthic communities and the relations between spatial and temporal variations. The resemblance of the latter to Walther's Law lends the model some credence, particularly to the paleontologist.

The conceptual model proposed here can be used to predict some of the biological consequences of man's activities. Probably one of the most drastic changes that can occur in shallow water environments is the alteration of the substrate. Dredging and filling operations, erection of submarine structures, overgrazing in the watershed of a bay or estuary often result in changing patterns of deposition and erosion. In addition to altering the texture of the sediment, such changes probably affect the quality and quantity of the food of the detritus feeders. If the local order of succession can be reconstructed for several communities, then the model can be used to predict the order of events, when the substrate is altered, in terms of local species.

For example, the introduction of clay and silt-size particles into a clean sand area would be expected to downgrade the clean sand assemblage. The mud species low in the order of succession in muds would become more common among the most diverse sand species assemblages. These intergrading species would spread throughout the region. Eventually, the species characteristic of higher stages of succession in muds would appear in the muddiest parts of the area. These circumstances can be represented diagrammatically by viewing *figure 8–3* on end.

A similar approach can be used to predict the effects of other kinds of general disturbance. Obviously, a pre-disaster survey of the communities in the area under consideration is essential. Pollution of marine waters by domestic and some kinds of industrial sewage appears to have the effect of downgrading the entire community. While oil spillages may completely destroy a littoral community, we would expect it to recover in a successional fashion. The model could also be used to develop standards for controlled disturbance and recovery.

Predicting faunal changes in continuous stratigraphic sequences could be performed in a similar fashion. The order of succession within a relatively homogeneous stratigraphic unit could be reconstructed from an analysis of variations in species composition in many outcrops. Knowing the order of succession would enable the paleontologist to recognize anomalies and to relate communities in space and time. If the order of succession is known for two different communities, then it should be possible to predict the faunal changes as one grades horizontally or vertically into the other. By these means, model faunal sequences might be developed for alternative stratigraphic sequences.

The view of the nature of shallow water benthic marine communities taken here is that they are low-grade communities, largely controlled by the physical environment. This is not meant to imply that biological interactions are unimportant but only that they exert less control of the structure and organization in these kinds of communities than in others.

9

MODELS IN BIOGEOGRAPHY

Daniel Simberloff

Editorial introduction. To most invertebrate paleontologists, biogeography chiefly means plotting faunal provinces. In the past decade, however, biogeography has come into its own as a branch of experimental ecology, and it is this new emphasis which Simberloff discusses. In this essay, Simberloff, at our special request, has taken special pains to translate much of the biogeographical revolution of the 1960's into a form understandable to the typical invertebrate paleontologist—i.e., a person with little mathematical training. The meaning of the equations is explained.

If any reader wishes to consider the chapter without any reference to mathematics, he can do so, and the chapter is still understandable. The main point, that events in ecologic time predetermine results traditionally seen by paleontologists in geologic time, is clear. MacArthur (1972) also emphasizes this in a broad, basic book treating several ecological aspects of distribution.

The concept of a species equilibrium, which is a balance of immigration and extinction, was a major conceptual touchstone of this revolution. However, the role of evolutionary rates as a pacemaker, modifier, or merely a minor aspect of the species equilibrium is a problem yet to be analyzed. As Simberloff concludes, paleontologists are in a unique position to contribute in new ways to the understanding of experimental biogeography.

Introduction

During the past decade biogeography has undergone a radical shift in focus, as an entirely new approach to problems of geographic species distributions has developed. Traditionally, biogeographers have been interested in six major phenomena:

(1) The geographic origins of individual taxa, and dispersal routes away from these origins (Wallace, 1876; Merriam, 1894; Merriam *et al.*, 1910; Matthew, 1915; Willis, 1922; Darlington, 1957, 1965; Kendeigh, 1961).

(2) The evolutionary consequences of isolation, particularly on distant oceanic islands (Darwin, 1859; Lack, 1947; Darlington, 1957; Carlquist, 1965).

(3) Differing rates of evolution in subgroups of the same taxon located in different regions (Willis, 1922; Fischer, 1960; Simpson, 1964; Stehli, Douglas, and Newell, 1969).

(4) Different numbers of species in regions with different areas (Arrhenius, 1921, 1923; Gleason, 1922, 1925; Dony, 1963; Hamilton, Barth, and Rubinoff, 1964; Williams, 1964; Johnson, Mason, and Raven, 1968; Malyshev, 1969).

(5) The "biome" concept—the relationship between ecological succession and the geographic distribution of organisms (Clements and Shelford, 1939).

(6) The determination of present biogeographic distributions by climatic conditions, especially temperatures, at the margins of species ranges, and the possibility of inferring climatic conditions in the past from fossil ranges (Hutchins, 1947; Slaughter, 1968).

Models dealing with the first phenomenon, dispersal routes and geographic origins, are largely qualitative, rest heavily on chosen well-studied examples, and speculate on the relative importance of such long-term factors as land bridges, continental drift, and aerial dispersal as a rare but recurring event. Lately this matter has been treated more quantitatively, with all available data used in statistical attempts to determine the degree of affinity between biotas of different regions (Hagmeier and Stults, 1964; Peters, 1968; Howell, 1969; Tobler, Mielke, and Detwyler, 1970).

The evolutionary consequences of isolation, particularly degrees of endemism and adaptive radiation and the appearance of structurally bizarre forms presumably able to survive in the relative absence of competition, are inherently a qualitative and example-oriented matter. Recently endemicity has been treated numerically (Hamilton and Rubinoff, 1967; Greenslade, 1968a), but the emphasis, mostly anecdotal, remains on the static existence of certain forms on certain islands. Short-term ecological effects have not been treated except as they are preconditions for evolutionary events.

Although the third phenomenon (different evolutionary tempos), since it deals in rates, is inherently dynamic, it has until recently been treated as a purely long-term matter unrelated to short-term interactions among species which might affect their coexistence in the same geographic region. Some of the efforts have been highly quantitative (Willis, 1922; Stehli, Douglas, and Newell, 1969) and others qualitative and example-oriented (Fischer, 1960; Simpson, 1964).

The fourth question—what determines the numbers of species in different locations, particularly when latitudinal variation is factored out—has been dealt with statistically for fifty years, and large masses of data rather than well-studied small taxa have been involved because of the nature of the problem. Until the most recent efforts (Johnson and Raven, 1970), however, all studies have considered the problem a static one in that changes in species number were conceived of as occurring only on an evolutionary time scale; short-term changes were not treated.

Although the biome was explicitly the result of an ecological process–succession—the process was thought to occur rarely or even just once in most regions, and the resulting climax which characterized the biome was viewed as a static community persisting over evolutionary periods of time. Research from the biome viewpoint was qualitative and largely descriptive.

Finally, the relation of climatic conditions to species ranges has been examined from data on several carefully studied species both in the marine realm (Hutchins, 1947) and for terrestrial vertebrates (Slaughter, 1968). Range limitations were viewed as static over the short term, with modification occurring even more gradually than some secular changes in climate (Slaughter, 1968).

There were unifying trends in the treatment of these six problems, reflecting a common underlying conception of the nature of biogeographic change. Primarily, there was an emphasis on events occurring in "evolutionary time" (Pianka, 1966) and a tacit downgrading of events occurring in "ecological time." That is, over the relatively short lengths of time involved in ecological interactions like competitive exclusions, local extinctions caused by climatic catastrophes, and dispersal of organisms over commonly achieved distances, changes in biotic distributions were regarded as rare and usually minimal. If S_i represents the number of species in location i, and P_{ij} the measurement of the jth physical parameter for location i (where $j = 1, \ldots, n$; and the parameters may include not only intrinsic characteristics of a location, like area and maximum elevation, but also characteristics of its relationship with other locations, like distance between the location and some other location), then it was assumed that over the short term, or ecological time,

$$S_i = f(\{S_k\}_{k \neq i}, \{P_{ij}\}, \text{all } i \text{ and } j) \tag{1}$$

where f is some function (in the biome model, this equation would still hold once the original succession had occurred). Equation 1 states that *the number of species in a region is determined by the physical characteristics of the region, including the proximity of other regions, and the numbers of species in other regions.*

A secondary element common to many traditional biogeographic models was a tendency to be qualitative, and another frequent approach was to treat a single species or genus as a paradigm for all those of a higher taxon.

These three attitudes have been explicitly discarded recently in the evolution of a new conception of biogeographic distributions, with an associated set of questions only partially overlapping those asked in the past.

The Equilibrium Model

Preston (1962) and MacArthur and Wilson (1963) proposed independently that the biota of an island at any point in time is a dynamic equilibrium created by a balance between immigration of species new to the island and extinction on the island of species already present. Thus, although species number would be approximately constant over short time periods (barring ecological change), biotic composition would vary continually because of local, short-term ecological interactions. Evolution was viewed as acting over geologic time intervals to modify the equilibrium gradually. Although the model was originally suggested for oceanic islands, and much of its subsequent development continues to be concerned with islands, it has been utilized as a general biogeographic scheme for regions as small as minute habitat patches (Levins, 1969) or as large as continents (Webb, 1969). Such applications will be discussed in a later section.

The equilibrium model, then, assumes that over ecological time, and within a successional stage,

$$\frac{dS_i}{dt} = g(\{S_i\}, \{P_{ij}\}, \text{all } i \text{ and } j) \qquad (2)$$

where g is some function, and a description of the state of a system of islands at any point in time (the values of the S_i's, which formerly had been assumed to be given by equation 1) would require solution of the system of differential equations 2, which says that *the rate of change in species number for any island is determined by the physical parameters of the region and the numbers of species in other regions*, and of a set of differential equations expressing the successional characteristics of each island as an isolated system (for instance, $dS_i/dt = h_i(S_i, t)$, where h_i is a function describing the successional characteristics of island i).

The function g has not been explicitly formulated and must be extremely complex, certainly more so than the most complex classical predation and competition equations since so many more species are involved and so many different sorts of single- and multi-species phenomena are important in determining what set of species will inhabit a given island at any point in time. But approximate solutions may be achieved if simplifying assumptions are made and if the numbers of species in the locations in the system other than the island of interest (the S_i's) are considered together as functions affecting the extinction and immigration rates of that island.

Let I_s = immigration rate of new species onto the island when S species are already present, in species/time.

E_s = extinction rate of species on the island when S species are present, in species/time.

P = total number of species in the species pool of the source area. (Source area for an island would be one or more mainland areas or other islands from which the species which colonize the island are drawn.)

All other factors remaining constant, I_s (the immigration curve) should be a monotonically decreasing function of S simply because the larger S (the more species already present on the island), the fewer species in the pool are left as potential immigrants. Furthermore, we might expect that the species in the pool are not all equally likely to immigrate to the island given that they are not already present. Rather, we would expect a distribution of immigration probabilities depending largely on dispersal capabilities and population sizes of the different species making up P. This problem will be dealt with explicitly later, but it can readily be seen that the immigration of the first ten species would be expected to lower the subsequent immigration rate more than the immigration of the second ten species, which in turn ought to lower the subsequent immigration rate more than the third ten species, etc. Consequently, the I_s curve will not only be monotonic decreasing, but ought also to have increasing (negative) slope.

E_s (the extinction curve), on the other hand, ought to be an increasing function of S for two reasons. First, the more species are present, the more can possibly be extinguished on the island from purely non-interactive causes like climatic catastrophes (bearing in mind that even in a climax community, the total number of species present at any point in time includes many transients from earlier stage communities which are doomed to quick extinction). Second, the greater the number of species that are present, the more frequently will competitive interactions cause exclusion. Since such interactions would be expected to be more intense the more species are present, one might also expect the slope of the E_s curve to increase. Although many interactions are believed to increase the number of coexisting species possible in a given environment through more effective resource partitioning, there is experimental evidence for at least one island ecosystem that competitive interactions are more significant numerically over the short term than those leading to an increase in species number (Simberloff, 1969). (Over evolutionary time periods, extinction rates may temporarily decrease as number of species remains constant or even increases. Discussion of this problem is deferred to a later section.)

These considerations regarding the shapes of the immigration and extinction curves are incorporated into *figure 9–1*. The immigration and extinction curves intersect at $S = \hat{S}$, and since at this point $I_s = E_s$, the same number

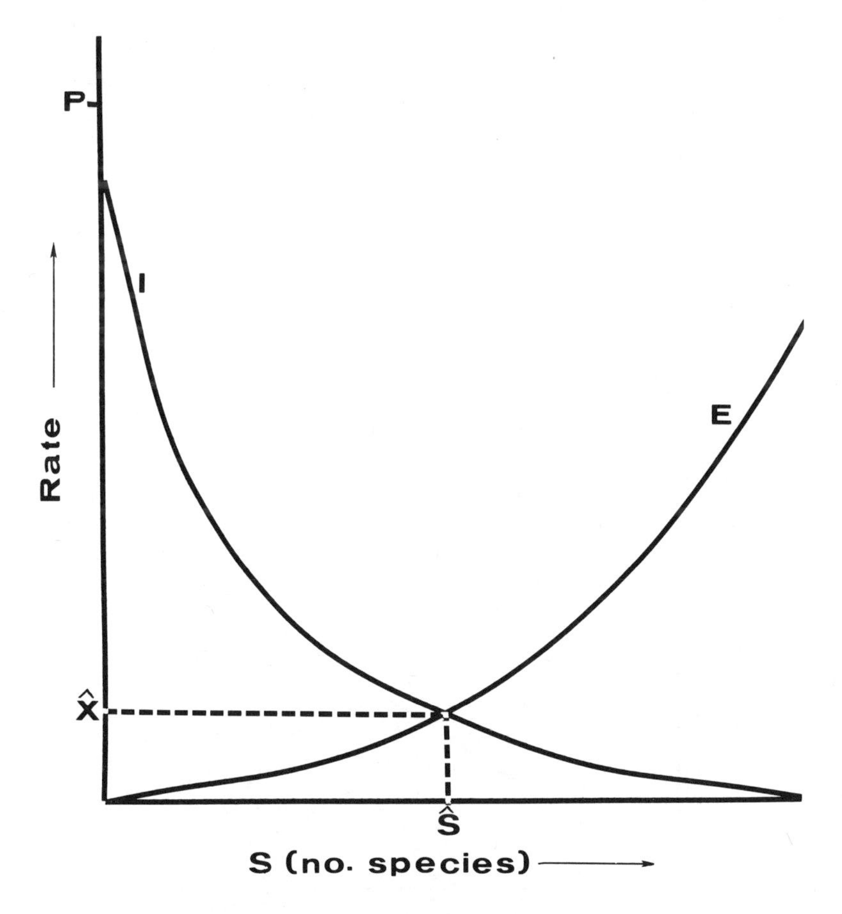

Figure 9–1: Simple equilibrium model for island biota. The equilibrium number of species $\hat{S}$ and turnover rate $\hat{X}$ are determined by intersecting immigration and extinction curves. (Adapted from MacArthur and Wilson, 1967; figure 7.)

of species immigrates as is extinguished and a dynamic equilibrium is maintained, $E_{\hat{S}}$, the extinction rate at equilibrium, is called the "turnover rate," $\hat{X}$, and $\hat{S}$ the "equilibrium number of species."

Further assumptions about the shapes of the immigration and extinction curves lead to interesting qualitative generalizations. We first simplify by making these curves linear, though the results would hold for curves of the general probable shape (*figure 9–1*) except as indicated. If two islands are identical in all physical characteristics except distance from source area, the extinction curves ought to be identical but we would expect I_s to be lower for all S on the more distant island because of the lower probability of any potential immigrants successfully traversing the greater distance. This results in a lower equilibrium number of species on the more distant island, a phenomenon well known to biologists from field data and called the *distance effect* (*figure 9–2*). An additional prediction is that turnover rate is higher on near islands, all other things being equal.

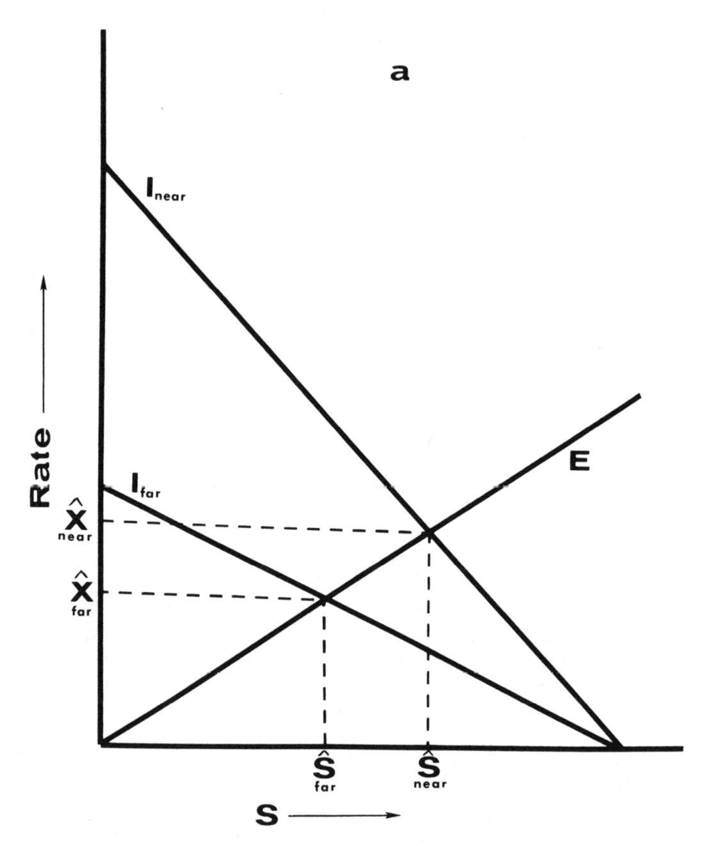

Figure 9–2: Equilibrium model interpretations of ecological phenomena of distance effect. More distant islands have fewer species at equilibrium.

For two islands equally distant from the mainland but with different areas the immigration rates ought to be equal at all S. At least this would be true if the arcs delimited by the two islands on the source area were equal, since this parameter must largely determine how many dispersing individuals of many species will land on an island. But one might expect E_s to be greater on the small island, first because competitive interactions might be expected to be more intense among a given number (S) of species the smaller the island, and second because population sizes will tend to be smaller on smaller islands and this may significantly increase the risk of extinction from chance climatic catastrophes. Consequently smaller islands will have fewer species than large ones, all other things being equal (*figure 9–3*), an observation made independently of the equilibrium model by many biologists. Smaller islands should also have higher turnover rates.

In addition to rationalizing previously known phenomena, the equilibrium model generates novel hypotheses which may be tested by statistical or even

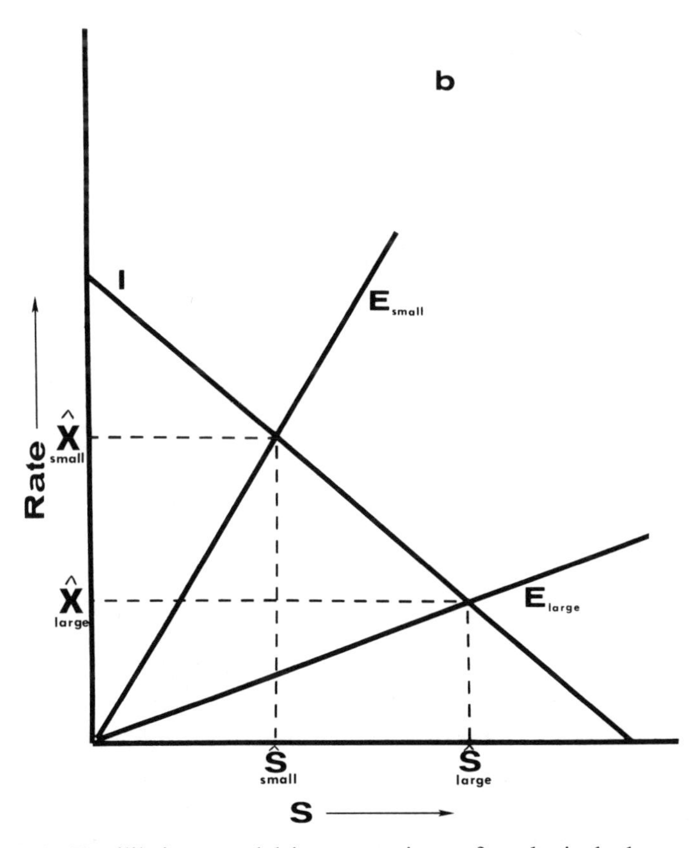

Figure 9–3: Equilibrium model interpretations of ecological phenomena of area effect. Smaller islands have fewer species at equilibrium.

experimental means (see below). Perhaps the most interesting is derived by combining the curves of *figures 9–2* and *9–3*, and is valid not only for linear I and E curves, but whenever the two curves are mirror images of one another, as in *figure 9–1* (MacArthur and Wilson, 1967). In *figure 9–4*, E_p is the extinction rate for the island in question when all P species are present and I_o the immigration rate when no species are present. By similar triangles

$$\frac{\hat{X}}{P - \hat{S}} = \frac{I_o}{P} \quad \text{and} \quad \frac{\hat{X}}{\hat{S}} = \frac{E_p}{P}$$

If both equations are solved for $\hat{X}P$ and the two solutions set equal, then

$$E_p\hat{S} = I_o(P - \hat{S})$$

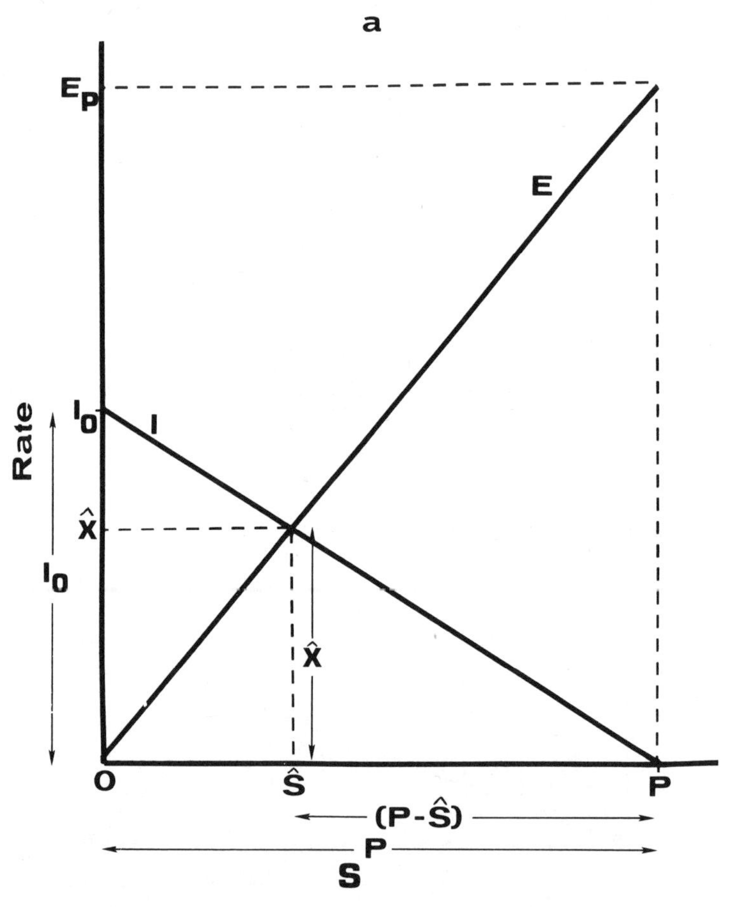

Figure 9–4: E_p is the extinction rate when all P species are present, and I_o is the immigration rate when zero species are present on the island. (Adapted from MacArthur and Wilson, 1967; figure 11b.)

which solved for $\hat{S}$ is

$$\hat{S} = \frac{I_o P}{E_p + I_o} \tag{3}$$

If distance is kept constant (I_o constant) and we change area, A (or maximum extinction rate E_p, which varies with area),

$$\frac{d\hat{S}}{dE_p} = \frac{-I_o P}{(E_p + I_o)^2}$$

or

$$\frac{d \log \hat{S}}{dA} = \frac{1}{\hat{S}} \cdot \frac{d\hat{S}}{dA} = \frac{1}{\hat{S}} \left(\frac{d\hat{S}}{dE_p} \cdot \frac{dE_p}{dA} \right) = \frac{E_p + 1_o}{I_o P} \left(\frac{-I_o P}{(E_p + I_o)^2} \cdot \frac{dE_p}{dA} \right) \tag{4}$$

$$= \frac{-dE_p/dA}{E_p + I_o}$$

But $-dE_p/dA$ is positive, so equation 4 says that a decrease in I_o (increase in distance) increases the rate at which species are added with increased area, or "log $\hat{S}$ [*and therefore number of species*] *varies faster with area on distant islands*." This is explicitly shown by differentiating equation 4 with respect to distance, D:

$$\frac{d}{dD} \cdot \left(\frac{d \log \hat{S}}{dA} \right) = \frac{(dI_o/dD)(dE_p/dA)}{(E_p + I_o)^2} > 0 \tag{5}$$

and since the order of differentiation may be changed in equation 5, MacArthur and Wilson simultaneously showed that "log $\hat{S}$ [*and therefore number of species*] *varies faster with distance on small islands*."

Whitehead and Jones (1969) relax the assumption that immigration rate is independent of area, and demonstrate that the above qualitative result still holds essentially because any effect of area on immigration rate must be monotonic increasing no matter what the distance. This sophistication undoubtedly makes the model more realistic; for example, one could imagine high-flying birds or visually orienting insects selecting the larger of two equidistant islands. And even non-visual species (seeds) could contribute to a higher immigration rate on larger islands, if a fixed number of individuals landed per unit time per unit area, land or water.

This result is pictured in *figure 9–5*. MacArthur and Wilson also derive an equation in the form of equation 2 by treating the I_s and E_s functions as implicit functions of the S_i's of the mainland areas and other islands and of the P_{ij}'s. The equation, applicable in a short-term situation (such as colonization of a volcanic island like Surtsey or Krakatau) where ecological phenomena are far more important numerically than evolutionary ones, is

$$\frac{dS(t)}{dt} = G[\hat{S} - S(t)] \tag{6}$$

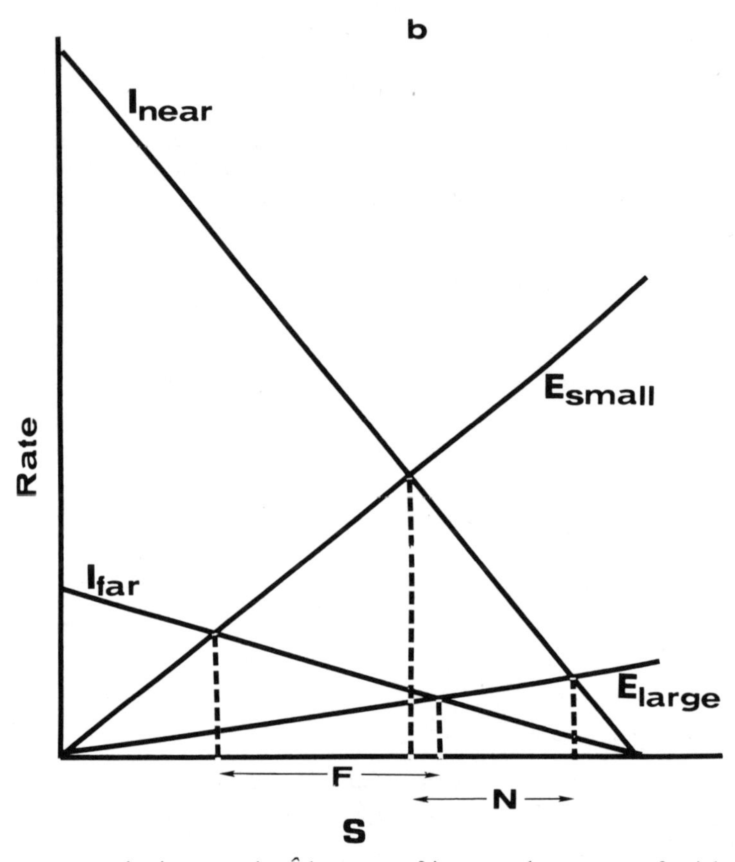

Figure 9–5: *F*, the increase in $\hat{S}$ because of increased area on a far island, is greater than *N*, the increase in $\hat{S}$ caused by an equivalent area increase on a near island.

where G is a constant and $S(t)$ is the number of species present on the island at time t. Equation 6 states that *the rate of increase in species number is proportional to how far the island is from an equilibrium condition*, and can be solved to yield

$$S(t) = \hat{S}(1 - e^{-Gt}) \tag{7}$$

which is plotted in *figure 9–6* and states that *species number increases monotonically with time, but at an exponentially decreasing rate, approaching the equilibrium number* ($\hat{S}$) *asymptotically.*

Equation 7 is then solved for the useful measure of the expected amount of time for a new or sterilized island to achieve a given fraction of $\hat{S}$, its equilibrium number of species, say 90%:

$$t_{0.90} = \frac{2.303}{G} \tag{8}$$

MacArthur and Wilson then deduce that the ratio of the variance of species number $S(t)$ to the mean of $S(t)$ for a set of identical islands falls from about 1 early in the course of colonization to about $\frac{1}{2}$ near equilibrium. Finally, equation 8 is put in a form which could conceivably be tested with available data:

$$t_{0.90} = 1.1515 \left(\frac{\hat{S}}{\hat{X}}\right) \tag{9}$$

A conceptual problem with this original equilibrium model is noted by Simberloff (1969). The immigration and extinction rates were assumed to be functions of species number S, when in fact this is at best a crude approximation. For E_s to be a single-valued function of S, every conceivable set of S species on the island would have to produce the same extinction rate, which is not true. And an average E_s for all $\binom{P}{S}$ sets of S species would have to

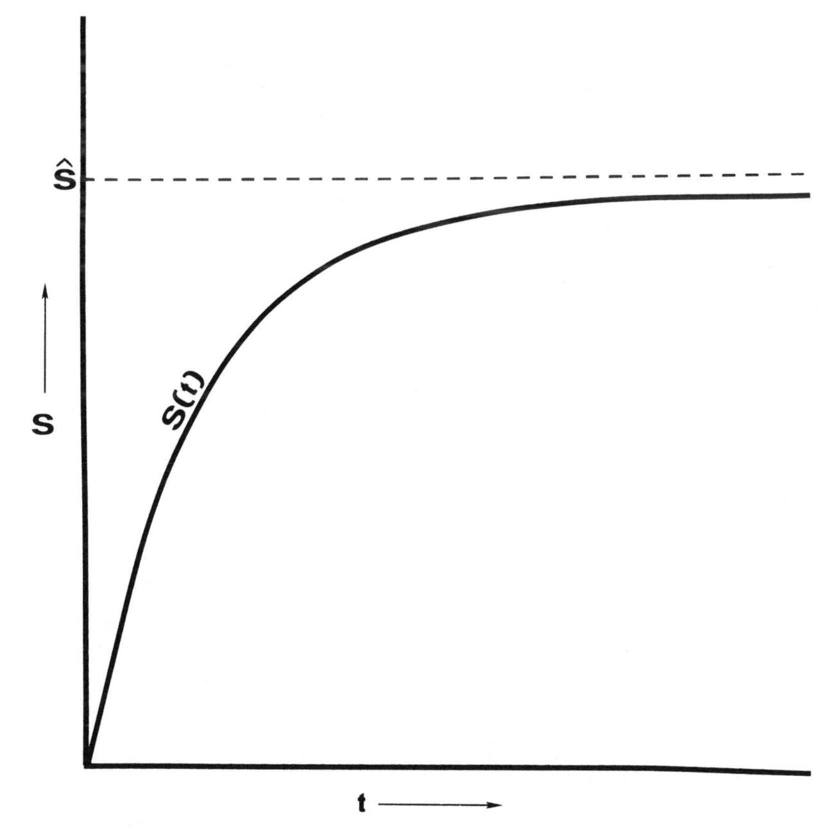

Figure 9–6: Colonization curve of expected number of species on an island vs. time: $S(t) = S(1 - e^{-Gt})$. (Adapted from MacArthur and Wilson, 1967; figure 20b.)

have each set weighted by some estimate of its probable occurrence. But even more serious is the failure to account for the distribution of the population sizes of the S species, which surely affects extinction rate at least as much as number of species does.

Immigration rates are usually not affected by the population sizes of the species already on the island, but certainly are affected by *which* S species are present on the island (and therefore which P-S species are left to immigrate).

A more general model which avoids this problem, and of which the MacArthur-Wilson model is a special case, was derived from characteristics of the species in the pool as well as the island in question (Holland, 1968; Simberloff, 1969). If we define a "propagule" of a species as the minimum number of individuals of that species capable of initial population increase under ideal conditions for that species (unlimited food, suitable habitat, etc.), let

i_α = invasion rate of propagules of species α falling on island in question, in propagules/time.

e_α = "species extinction" rate for species α of extinctions from non-interactive causes, if after every extinction a propagule of species α immediately invades. Units are extinctions/time.

It can be shown that this model produces an equilibrium number of species in the total absence of interactions among species, and an equation analogous to equation 3 can be deduced for that non-interactive equilibrium ($\hat{S}_n$):

$$\hat{S}_n = \sum_{\alpha=1}^{P} \frac{i_\alpha}{i_\alpha + e_\alpha} \tag{10}$$

A set of differential equations of the form of equation 2 can be derived, and their solution is

$$S(t) = \sum_{\alpha=1}^{P} \frac{i_\alpha}{i_\alpha + e_\alpha}(1 - e^{-(i_\alpha + e_\alpha)t}) \tag{11}$$

The similarity of this equation to MacArthur and Wilson's equation 7 is apparent. The right-hand member of equation 11 is a sum of P terms each of which is identical in form to the right-hand member of equation 7. Consequently the entire sum has a shape like that of *figure 9–6*, though only for special sets of i_α and e_α does equation 11 reduce to equation 7.

Formally, it can be shown that equation 7 is simply the special case of equation 11 where all species have the same i_α and e_α. For in deriving equation 7, MacArthur and Wilson assumed that I_s and E_s are "locally straight" for S near $\hat{S}$, which is equivalent to assuming no interactions among the species and that each species contributes equally to immigration and extinction rates.

A more general relationship between variance of $S(t)$ and mean $S(t)$ for a set of identical islands is also given by the non-interactive model, namely that the ratio of variance to mean falls from 1 at time 0 to approximately

$$\sum_{\alpha} \frac{i_\alpha}{i_\alpha + e_\alpha}\left(1 - \frac{i_\alpha}{i_\alpha + e_\alpha}\right) \Big/ \sum \frac{i_\alpha}{i_\alpha + e_\alpha}$$

near equilibrium. A 95% confidence band about $S(t)$ is calculated which accounts for the stochastic nature of the colonization curve. An expression comparable to equation 9 for expected time until 90% of the equilibrium $\hat{S}$ is reached was derived involving only the species characteristics $\{i_\alpha\}$ and $\{e_\alpha\}$, and the additional prediction was generated that $t_{0.90}$ (or t_x for any percentage x) will be lower on nearer islands (*figure 9–7*), which is equivalent to saying that more time is required for distant islands to reach saturation.

Even a new or sterile island eventually reaches a point where biotic interactions like competition and predation occur, and the equilibrium model here becomes more qualitative. Experimental evidence indicates that a dynamic equilibrium number of species still exists for each island, and that it is somewhat lower than the non-interactive equilibrium derived above (Simberloff and Wilson, 1969). To predict just how much lower would require much more information than is now available on interactions between species within a community and on higher order interactions. It seems possible that the community matrix approach of Levins (1968) to the problem of species number in a stable community could be merged with additional information on dispersal and non-interactive extinction characteristics of the species of an ecosystem to provide at least a theoretical model for interactive equilibrium.

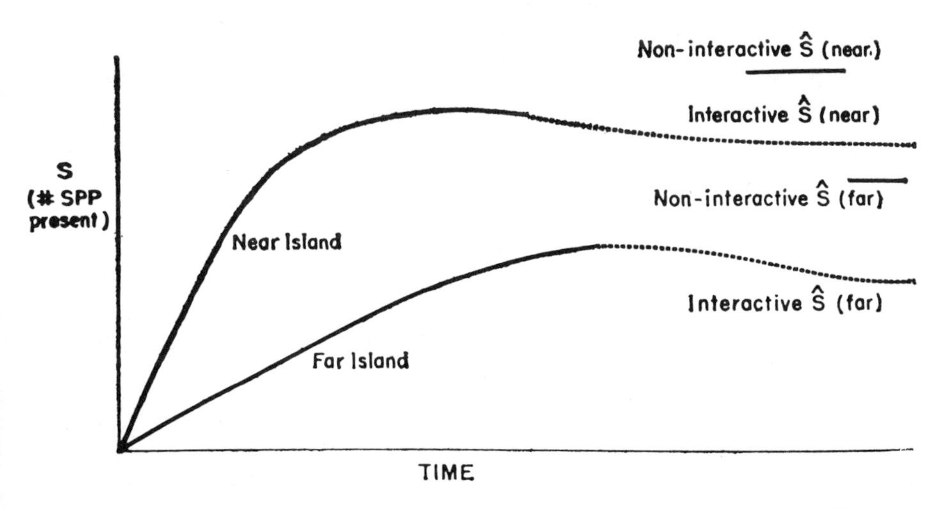

Figure 9–7: Effect of varying distance on colonization curve and equilibria. (Adapted from Simberloff, 1969; figure 2.)

Data from oceanic islands to test the equilibrium theory were scarce, with the anecdotal evidence from post-eruption colonization of Krakatau probably the best. These and similar data were all consistent with the notion of an equilibrium theory but provided too few reliable data points to substantiate it.

Recently an experiment was performed which remedied these shortcomings (Wilson and Simberloff, 1969; Simberloff and Wilson, 1969, 1970). A series of seven small red mangrove islands in Florida Bay, at various distances from the main Florida Keys, were censused exhaustively. The animal communities of these islands consisted almost entirely of arboreal arthropods, especially insects, and number between 20 and 50 on each island. The single plant species was the red mangrove which itself formed the island. Simplicity of the habitat, including the absence of supratidal ground, allowed complete censuses to be made with assurance, at least with respect to presence or absence and therefore number of species.

Control islands were also censused, and then all animal life on the experimental islands was removed without damage to the mangrove by methyl bromide fumigation. One island was "defaunated" six months before the others to determine if the recolonization of the islands had a strong seasonal component. All islands were censused regularly for one year after defaunation and irregularly after that. The following observations confirmed the general correctness of the dynamic equilibrium model:

First, on all islands but the most distant the number of species rose to a point slightly above the pre-defaunation number, then fell and oscillated in the vicinity of this figure. The interpretation is that S rose to a point above the longer range interactive equilibrium *en route* to the higher non-interactive $\hat{S}$ (*figure 9–7*). The control islands did not show significant or consistent change in species number.

Second, distant islands had lower equilibria and took longer to return to the vicinity of the original $\hat{S}$ after defaunation. The most distant island has not yet reached the original number of species three years after treatment, and in fact on this island the characteristic overshoot may be eliminated because those few species which have colonized may be able to achieve large population sizes before their competitors and predators immigrate.

Third, the shape of the colonization curve of number of species vs. time conformed approximately to that predicted by equations 7 and 11, and $t_{0.90}$ calculated by equation 9 with an estimate of turnover rate $\hat{X}$ was within the range of the true value allowed by the errors of the approximations for $\hat{X}$, $\hat{S}$, and $t_{0.90}$.

Finally, the equilibrium was *dynamic*, with turnover rate $\hat{X}$ at equilibrium on an island 200 yards from nearest source area between one-half and one species/day, or approximately 2% of the fauna each day.

Further experiments on entire biotas are necessary, as are more systematic data on limited taxa from larger oceanic islands. A recent effort along the latter lines is by Diamond (1969) on birds of the California Channel Islands; hopefully the data from the new volcanic island of Surtsey will be even more comprehensive.

The equilibrium model is theoretically applicable to any clearly delimited community, including "habitat islands" such as ponds, bromeliads, mountaintops, etc. MacArthur and Wilson demonstrate that the results of Maguire's (1963) experiments on colonization of bottles of sterile water placed at various distances from freshwater ponds by aquatic microörganisms are consistent with the equilibrium model. Other attempts to use habitat islands to test the equilibrium model include those of Cairns *et al.* (1969) on colonization of plastic "islands" by aquatic protozoans, Patrick (1970) on the colonization of glass slides by diatoms, Culver (1970) on Appalachian caves as islands colonized by aquatic arthropods, and Hubbard (1971) on all organisms colonizing new artificial ponds. All these experiments confirm the notion of an equilibrium number of species for each "island," and to varying degrees other aspects of the model. However, most suffer from the limited nature of the biota examined.

Main areas where advances in the equilibrium theory are to be sought, then, are better data on dispersal and non-interactive extinction characteristics of individual species, elaboration and sophistication of the model where species interactions become important, and systematic data on more real island ecosystems, particularly on entire biotic communities.

The Importance of "Stepping Stones"

A complication of the basic equilibrium model dealt with in some detail by MacArthur and Wilson involves colonization of an island (or mainland area) not directly from a source area, but rather via "stepping stone" islands. Since in many parts of the world (especially the Pacific Ocean and Caribbean Sea) numerous small islands are interspersed between major land masses, this complication is probably a realistic one.

Qualitatively, if one had a distant island, a source area, and a small intermediate stepping stone, one might expect a high invasion rate of propagules on the distant island, but most of these belonging to the few species on the stepping stone. Consequently the I_s curve should be lower at all S for an island distance D from a stepping stone instead of distance D from the major source area itself. Furthermore, the difference between the two curves should increase with S, since the greater the number of species on the island, the fewer the species from the stepping stone which will be left to immigrate. Consequently the immigration curve should be steeper in stepping stone colonization, and the extinction curve unchanged from that in colonization directly from source. This may lead to a decreased effect of area on $\hat{S}$ (*figure 9–8*).

MacArthur and Wilson (1967) and Wilson and Hunt (1967) relate the use of stepping stones to the question of biotic exchange between two regions, first attempting to determine the relative proportions of propagules arriving on an island from a stepping stone and a more distant source, respectively. Assuming that the three islands are colinear, let

w_1 = width of stepping stone
w_2 = width of source
w_r = width of recipient island
d_1 = distance from stepping stone to recipient
d_2 = distance from source to recipient.

(All w_i's measured at right angles to the line joining the three locations.)

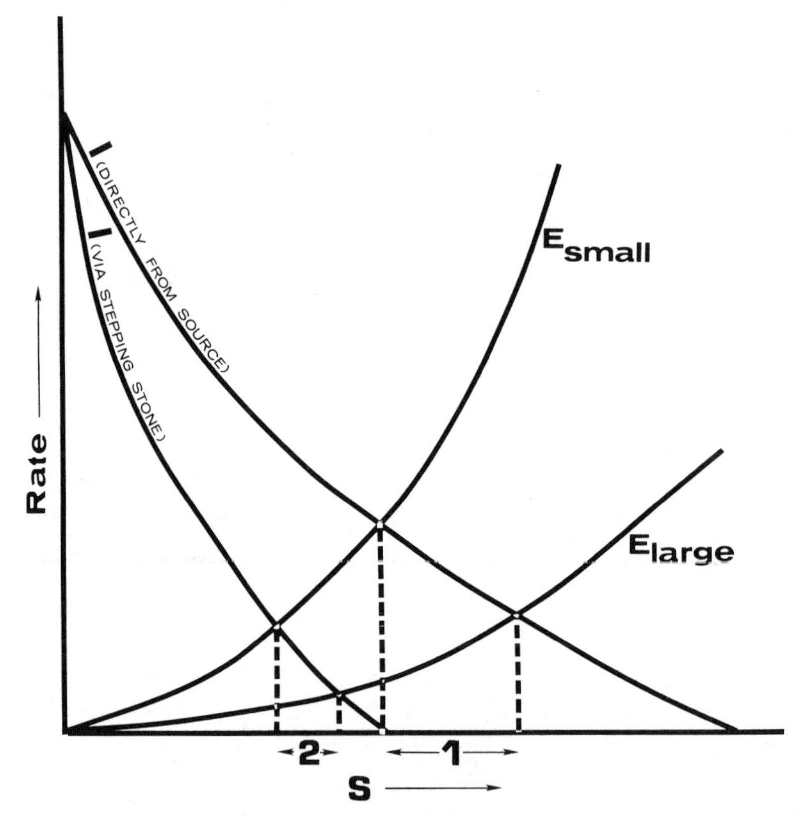

Figure 9–8: Effect of stepping stone on island colonization. 1 = immigration curve for island distance *D* from source area itself; 2 = immigration curve of island distance *D* from small stepping stone island. Note decreased area effect on $\hat{S}$: difference 1 > difference 2. (Adapted from MacArthur and Wilson, 1967; figure 14.)

The number of propagules of a given species leaving a location should be approximately proportional to the population size of the species at that location, which should be approximately proportional to the area of the location. Consequently the number of propagules of a species leaving location i/unit time should be αw_i^2, where α is a fitted constant. Most of these αw_i^2 propagules will perish, and the number reaching a given distance d depends on the dispersal distribution of the species in question, which is a function of the dispersal means.

One possible distribution would be the exponential distribution resulting if the probability of a propagule's perishing per unit time (or per unit distance, if rate of travel does not change) remains constant. Such a distribution might be expected in passive wind dispersal, for example. The number surviving after traveling distance d_i will be $\alpha w_i^2 \, e^{-d_i/\lambda}$, where λ is the mean distance traveled. In addition, we must also be concerned with whether the propagule took the right direction, and if no selective behavior, differential wind direction, etc., are involved, the probability of taking the right direction is

$$\frac{2 \tan^{-1} (w_r/2d_i)}{360^\circ}$$

Thus the total number of propagules reaching the recipient island from location i is

$$n_i = \frac{2 \tan^{-1} (w_r/2d_i)}{360^\circ} \alpha w_i^2 \, e^{-d_i/\lambda} \tag{12}$$

Another possible distribution would occur if the propagules actively fly to exhaustion, float, or are carried on a raft (all at constant speed and direction). We would then expect persistence time to be normally distributed, and it can be shown that

$$n_i = \frac{2 \tan^{-1} (w_r/2d_i)}{360^\circ} \alpha w_i^2 \left[1-2 \int_0^{d_i} \frac{1}{s\sqrt{2\pi}} \exp -\frac{1}{2} \left(\frac{x}{s}\right)^2 dx\right] \tag{13}$$

Because the most important term determining propagule attrition in both equations 12 and 13 is the exponential term, in which d_i^2 appears in normal disperson and d_i in exponential dispersion, the rate of attrition is much higher for normal dispersal, while for both types of dispersion the directional component is much less important than the exponential term.

Since we are interested in the *ratio* $n_1/(n_1 + n_2)$ rather than the absolute numbers of propagules, α, the most difficult term to measure, is eliminated. MacArthur and Wilson calculate isoclines of this ratio for both forms of dispersal (*figures 9–9* and *9–10*) and conclude that when stepping stones of even small size are near the recipient island, the majority of propagules are contributed by the stepping stone rather than by the source. This preponderance is greatest in normal dispersal, and for both kinds of dispersal the

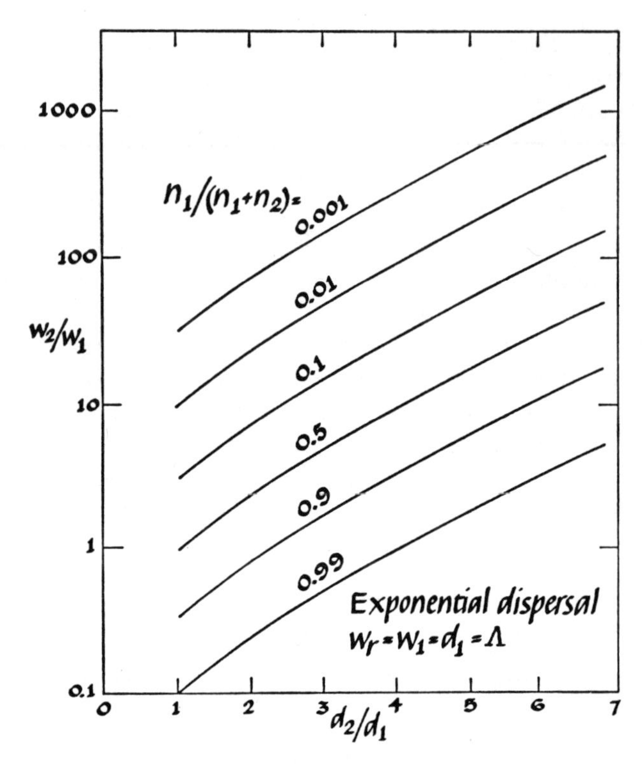

Figure 9–9: Exponential dispersal to recipient island *via* stepping stone island. Isoclines of $n_1/(n_1 + n_2)$ give the probability that a given propagule arriving on the recipient island came from the stepping stone (island 1) rather than the main source area (island 2). Symbols as in text. Note that even tiny stepping stones can contribute the majority of propagules if the distance from stepping stone to recipient island is short enough. (From MacArthur and Wilson, 1967; figure 44.)

stepping stone becomes overwhelmingly important when the mean dispersal distance is just a small fraction of the distances between locations.

Wilson and Hunt apply the above model to the problem of Samoan ants, and determine that by either type of dispersal, Melanesian ants are likely to have arrived on Samoa (and more distant Polynesia) from tiny Futuna and the Wallis Islands rather than directly from Fiji or other large Melanesian sources.

Certain radically mixed biotas may be explained by different dispersal distributions of different segments of the biotas. For example, the Florida Keys have predominantly Antillean insects and higher plants (which might be expected to disperse exponentially) and North American vertebrates (which possibly approximate a normal dispersion). The best-known such

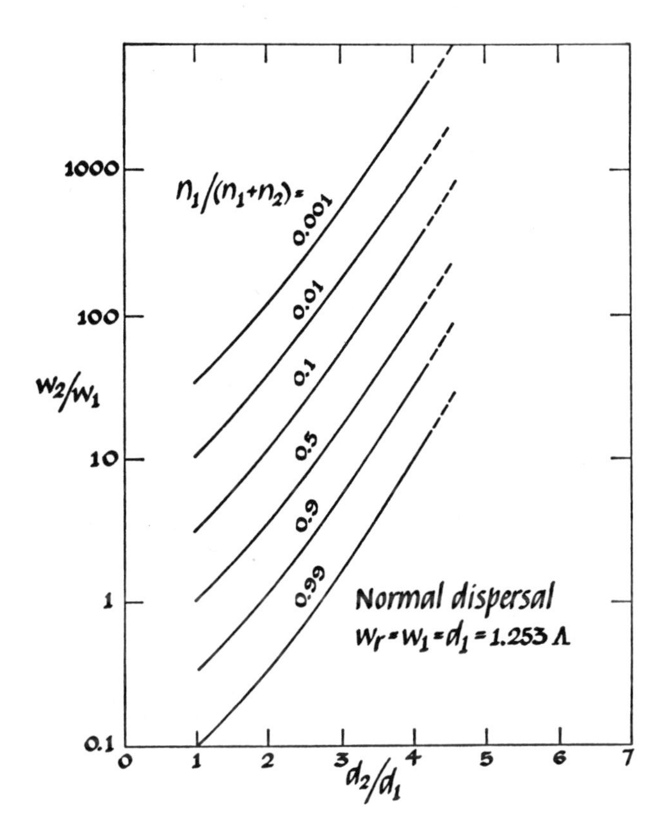

Figure 9–10: Normal dispersal to recipient island *via* stepping stone island, with conventions as in *figure 9–9*. Observe that stepping stones are even more important in normal dispersal than in exponential dispersal. (From MacArthur and Wilson, 1967; figure 45.)

area is New Guinea, with Oriental insects and higher plants and Australian vertebrates. MacArthur and Wilson point out that, depending on the as yet poorly known parameters of mean dispersal distances, exponentially dispersed insects and plants and normally dispersed vertebrates could lead to the observed pattern (*figure 9–11*).

As with the entire equilibrium model, further development and testing of the hypotheses on stepping stones require better data on dispersal characteristics of most of the species in a given ecosystem.

Further Consequences of the Equilibrium Model in Biogeography

Among others, the following two examples illustrate the shift induced by the equilibrium theory from the conception of biogeographic distributions

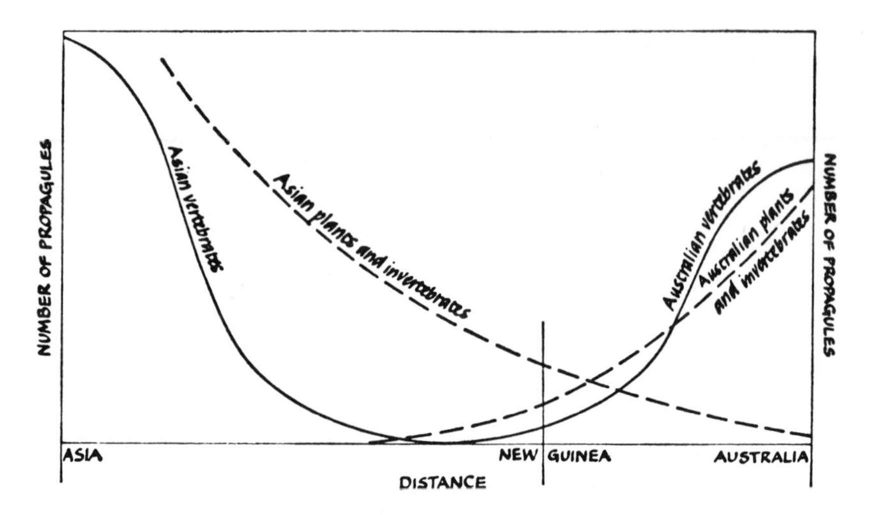

Figure 9–11: Possible relation between differences in dispersal distributions and origins of mixed biota. New Guinea is closer to Australia than to Asia, but receives more plant and invertebrate propagules from Asia because their exponential dispersal allows the greater number of propagules on Asia to compensate for the greater distance they must travel. But vertebrates, dispersing normally, primarily arrive from Australia; almost all Asian propagules perish at sea because of the greater distance. (From MacArthur and Wilson, 1967; figure 51.)

as largely evolutionary phenomena to the notion that they are more directly the result of short-term ecological events.

The relationship between species number and area has been a central problem in biogeography throughout this century (references in Introduction). It has long been noticed (Johnson and Raven, 1970, for references) that for most taxa and most regions the following equation holds:

$$S = kA^z \tag{14}$$

where S is the number of species (presumably at equilibrium), A the area of an island or continental region, k a fitted constant, and z a constant with a value in the vicinity of 0.3. This states that *if species number is plotted against area for a set of similar islands in a log-log plot, a straight line with slope around 0.3 results.*

In attempts to explain this empirical finding, simple curve-fitting evolved into more complex curve-fitting, exemplified by recent complicated multiple regression analyses with various transformed parameters. The only major deviation from this trend has been the model of Preston (1962), which views species-area relationships as a consequence of a canonical log-normal species-individuals distribution; the chief justification for the log-normal species abundance curve remains closeness of fit to published data points, however.

The method of approaching the species-area problem in the light of the equilibrium model has remained largely unchanged—multiple regression—but the interpretation of results is different (Greenslade, 1968a, b; Johnson and Raven, 1970; Vuilleumier, 1970). Whereas past emphasis had been on the static fact of a given number of species in a given location, current work regards the species set as a percent "saturation" of the equilibrium which would be observed if the island were adjacent to the source, the difference between the two values being due to lowered immigration rate with increased distance. This accords with the observation that distant islands of an otherwise similar set of islands frequently fall below the line fitted to equation 14.

A consequence of viewing extinction as a rare event, occurring in evolutionary time, was that the observed increase in species number with increasing area was attributed not to the effect of area *per se* but rather to the fact that habitat diversity, believed to be the major determinant of species number, increased with area (Hamilton, Barth, and Rubinoff, 1964; Watson, 1964; Williams, 1964; Johnson, Mason, and Raven, 1968). This conception is epitomized by Whitehead and Jones (1969), who assert that an observed increase in slope at 3.5 acres in the species-area curve for the Kapingamarangi atoll flora is due to the addition of a new habitat (fresh water) at that size. Although many of the species which appear on islands greater than 3.5 acres in this ecosystem are shown to require fresh water, it seems that area itself may be contributing to the increased species-area slope as follows.

If we assume that population sizes are roughly proportional to area, and if, as MacArthur and Wilson (1967) claim, every species has a critical population size below which extinction is very rapid (at least an order of magnitude more so than for populations above this size), one would expect smaller islands to have higher extinction rates irrespective of habitat diversity simply because more of their species will have population sizes below their critical sizes. But higher extinction curves, and equal or lower immigration curves, will mean that small islands have fewer species because of area alone (*figure 9–3*) in addition to whatever decrease is due to fewer habitats.

A recent unpublished experiment by Simberloff on red mangrove islands similar to those which were defaunated confirms the above reasoning. Eight islands were exhaustively censused for animal species number, after which various fractions of the islands were cut off to below the high tide level. After a period for equilibration the islands were censused again, while a control island was censused before and after the treatment. Number of species on the control island was unchanged, while all the islands with decreased area showed small declines in species number (*ca.* 5–10% of the original $\hat{S}$). Since no microhabitats were removed, and in fact the microhabitat proportions were virtually unchanged, area apparently had an independent effect on species number. A continuation of this work will attempt to confirm the relation between the area effect and distance posited earlier, and to construct species-area curves for single islands, thereby obviating regression analyses.

Another phenomenon which had been treated evolutionarily until recently is geographic variation in genus size. Mayr (1963) considers the number of species in a genus to be determined over evolutionary time by a balance between two opposing long-term forces: speciation, which leads to large genera, and adaptive radiation, which leads to smaller, even monotypic ones if there is little associated speciation. Grant (1966, 1968) and Simberloff (1970) have examined the mean species per genus ratio on islands from an ecological rather than evolutionary point of view. The rationale for this new approach involves the recent evidence on high avifaunal turnover rates (Mayr, 1965a; Diamond, 1969). For if island-wide extinctions are common rather than rare events, one would expect that such ecological phenomena as interspecific competition might have contributed to the present mean species per genus ratio on any island by having operated differentially between congeners as opposed to between more distantly related species at some time in the past.

Grant noticed that generally the mean number of species per genus on an island is lower than on its presumptive source area, and that for islands similar in other respects mean number of species per genus is lower on small islands than on large ones. He also determined that a subsection of a mainland source equal in area to an island not only has more species than the island, but also a higher mean species-to-genus ratio.

These observations are united in an hypothesis stating first that resources tend to be less diverse on islands, with least diversity on the smallest islands. This means that coexistence of species with similar ecological requirements is more difficult on islands than in mainland regions, the difficulty increasing with decreasing island size. Since congeners would tend to have more similar ecological requirements than more distantly related species, coexistence of congeners on islands is rarer than on the mainland. Other phenomena claimed to be associated with this resource paucity on islands are morphological and population adaptations among congeneric species, but it is the rarer coexistence *per se* which leads directly to the lower observed mean number of species per genus.

Simberloff has determined by computer simulation that the *expected* mean number of species per genus for subsets of species drawn randomly from the same species pool is a nearly linear monotonic decreasing function of the number of species drawn (*figure 9–12*) and that this is sufficient to explain all the observations of the last paragraph. An approximation to the expected mean number of species per genus for a subset of size S is

$$\frac{(S-1)[(\sigma/\gamma)-1]}{\sigma-1}+1$$

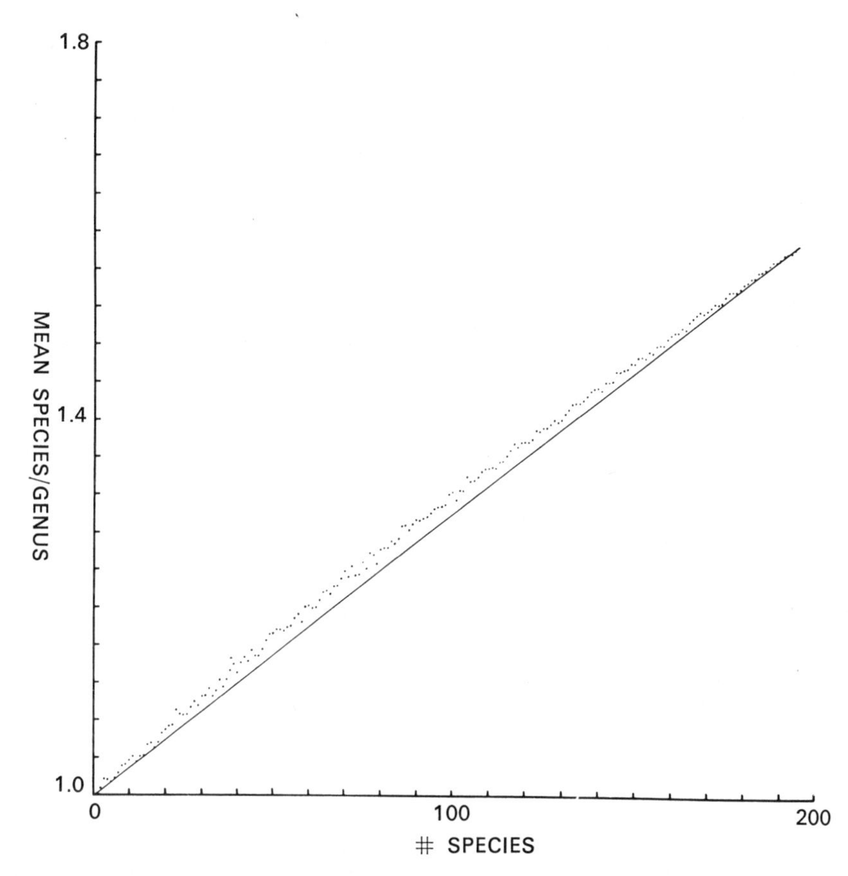

Figure 9–12: Plot of expected mean number of species per genus vs. number of species for subsets of land birds of Morocco drawn "randomly" by computer. The line is the straight-line approximation described in the text. Notice that the smaller subsets would be expected to have fewer species per genus. (From Simberloff, 1970; figure 1.)

where σ = the number of species in the source biota, and γ = the number of genera in the source biota.

In fact, a close examination of the mean species-to-genus ratio for land birds and for plants on a number of islands indicates a statistically significant tendency to have slightly *more* species per genus than expected on a chance basis. Simberloff interprets this as a consequence of the similarity of both dispersal characteristics and ecological requirements of congeneric species. Such similarities apparently outweigh whatever increased difficulties of coexistence are caused by a tendency to greater interspecific competition, a claim partly made by Williams (1964) as well.

The basic equilibrium model was clearly concerned with numbers of species rather than which ones and made no predictions about such matters as how many species might be expected to be held in common by two islands or continental regions, or which species drop out as area decreases for a series of islands. The work of Grant and Simberloff just described, along with preliminary work by Schoener (1965, 1968), Levins (1968), and others on characteristics of individual species and their relation to probability of successful island colonization, represents a first attempt at incorporating the "Who" into the equilibrium theory in addition to the "How many."

An example of current work in this area concerns the problem of the total number of equilibrium species for a single large island versus several small ones (an archipelago) with aggregate area equal to the large one. The situation at the poles of the earth might be construed as falling in this category. As an example, consider two similar small islands (say, x and y) and one large one (z) with area equal to the sum of the two small areas. One might ask whether there will be more or fewer species on x and y together than on z, and intuitively see two contradictory possibilities. If only a small fraction of the species pool consists of species adapted for successful colonization of small islands one might expect islands x and y to have virtually the same species (namely, these), whereas island z may have these plus a number of others which can survive on a larger island. This could be true even if the larger island had no additional habitats so long as it was sufficiently larger than island x or y to allow a number of species to raise their population sizes above their respective "critical" extinction levels postulated earlier in this section. On the other hand, if most species in the pool are highly and equally adept at surviving on small islands like x and y, one might predict that, by chance, x and y will have some pairs of different species (say, congeners) which would be unable to coexist on the same island even if the island were larger (like z). One might then expect x and y together to have more species than z.

Formally, MacArthur and Wilson suggest that if dispersal abilities alone are considered, and $\hat{S}$ is the equilibrium of species on x alone, the number on x and y together at equilibrium will fall somewhere between $\hat{S}_x$ (where there is a strong, well-defined gradient or order of dispersal capabilities among the P species in the pool) and $[2 - (\hat{S}_x/P)]\hat{S}_x$ (where all species disperse equally well). Above I have suggested that when competitive exclusions and other characteristics of the individual species in the particular pool are considered, the number of species on x and y together may be even higher than MacArthur and Wilson's upper limit, possibly approaching $2\hat{S}_x$.

An expanded equilibrium theory, then, should predict an equilibrium number of species for x and y together between $\hat{S}_x$ and $2\hat{S}_x$, and this would be compared to the $\hat{S}_z$, which would be predicted by equation 14 from the known constant $\hat{S}_x$ and the areas of x and z. An empirical test of this admittedly sketchy theory is currently under way on Florida Keys mangrove islands. Several large islands are being exhaustively censused, after which narrow channels will be dug through them, resulting in archipelagoes of various numbers of islands virtually identical in total area (and other characteristics like distance and microhabitat diversity) with the original single large islands. After equilibration, the islands will again be censused and the numbers of species observed will answer the original question of whether one large island should have more or fewer species than two smaller ones.

The Role of Evolution in the Equilibrium Model

The development of the equilibrium model involved the tacit assumption that ecological events are at least as important as evolutionary ones in determining biotic distributions. This was diametrically opposed to the previous approaches, as outlined in the introduction, and in the light of real-island turnover rate data discussed above the opposition seems justified. Clearly, though, speciation and global (rather than local) extinction are important determinants of species number on islands and continents, and we must now relate them to the equilibrium model.

Webb (1969) suggests that over millions of years even continental biotas are in approximate equilibrium, with speciation balanced approximately by extinction. Documenting this argument with data on North American mammals, he points out that, even though gross long-term climatic changes may have shifted the E_s and/or I_s curves (in this context, I_s stands for origination, which equals immigration plus speciation), over most periods of time the number of genera remained remarkably stable. Where changes in the extinction and/or origination curve did occur, they were not so abrupt as is commonly believed, and different periods of increase in number of genera have been due to both lowered extinction and raised origination curves, respectively. Turnover rates for the mammalian fauna are estimated at 1 species every 10,000–20,000 years, several orders of magnitude lower than for the arthropods on small mangrove islands and for birds on several island groups. Undoubtedly a major cause of this difference is the much larger area of continents, which greatly lessens the chance of extinction of all populations of a species by either interactive or non-interactive factors.

Just as the equilibrium model over ecological time was successfully applied to small "habitat" islands as described earlier, it seems possible that an equilibrium between origination and extinction such as that proposed by Webb might pertain over evolutionary time for such larger distinct regions as continents, marine faunal provinces, and even the earth itself. Since this sort of equilibrium cannot be confirmed by direct experiment, testing of appropriate data against various formal consequences of the equilibrium model will be necessary—what are the relationships between species number and area, turnover rate and area, turnover rate and distance, etc.?

Evolution not involving speciation can occur over much shorter intervals of time, such phenomena as character displacement and ecological release being possible over just a few generations (MacArthur and Wilson, 1967). Wilson (1969) views the colonization curve over evolutionary time as gradually rising after the initial interactive "quasi-equilibrium" described above is reached. First, over long periods of ecological time, continual turnover will result in combinations of species which are more highly co-adapted to one another and therefore allow the entire combination to persist longer. Eventually a very stable combination appears which persists indefinitely in ecological time, which Wilson calls the "assortative equilibrium," and which is analogous to the "climax" community of classical plant ecology. The number of species in this equilibrium combination is probably greater than in the earlier interactive "quasi-equilibrium" because one component of the increased co-adaptation is likely to be more efficient partitioning of resources.

Over short periods of evolutionary time, one might expect both the mutual co-adaptation of the species and the adaptations of the individual species to the local physical environment to become finer because of slow genetic changes in all species. Consequently the E_s curve will be lowered, while the I_s curve should not be greatly affected; $\hat{S}$ will therefore rise to an "evolutionary equilibrium," presumably maintained by speciation and extinction as discussed by Webb. This entire process is pictured in *figure 9–13*.

An estimate of just how much greater the evolutionary equilibrium will be than the first interactive or assortative equilibria is derived by Wilson and Taylor (1967) for the ant faunas of Polynesian islands east of Samoa. Observing that these islands have few if any native ant species, but rather various subsets of 35 "tramp" species dispersed by human commerce, and that these subsets apparently represent assortative equilibria, they then compare species numbers on these islands with those on Samoan islands containing apparently highly co-adapted native faunas. For an island of 430 square miles, the evolutionary equilibrium is shown to be 1.5–2 times an early interactive quasi-equilibrium. So long as the slope of the species-area curve is the same for evolutionary equilibria as for interactive equilibria, an estimate of the expected increase caused by evolution can be readily derived for an island of any size.

In addition to allowing a gradual increase in $\hat{S}$ over long periods of time, evolutionary changes in individual taxa may ultimately determine the dispersal rates and other ecological characteristics which are the driving force behind the dynamic ecological equilibrium. Wilson (1959, 1961) has proposed a "taxon cycle" model of Melanesian ant distribution. These ants are Asian in origin, and undergo the following characteristic changes. First, widespread

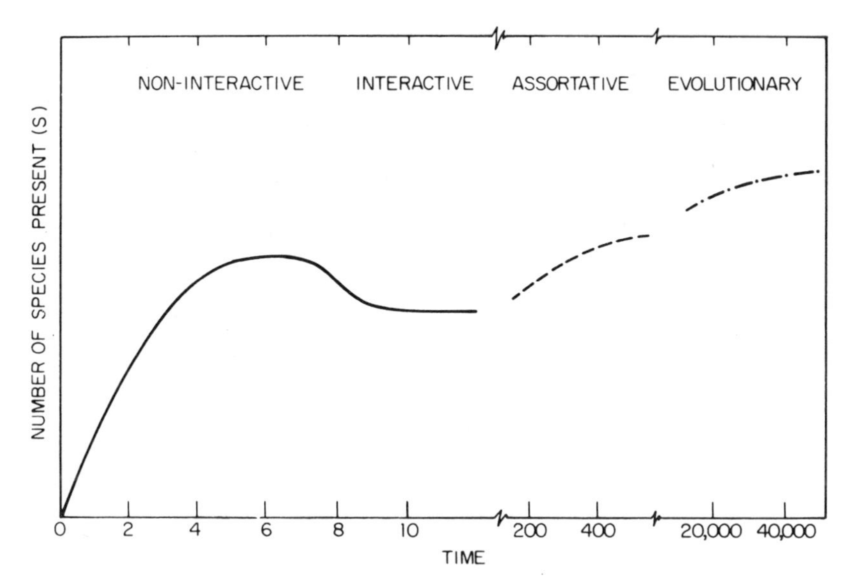

Figure 9–13: Sequence of "equilibria" on an island colonized from a sterile condition. The time scale is relative. (From Wilson, 1969; figure 5.)

species in southeastern Asia adapt to marginal habitats, particularly the littoral habitat from which transoceanic dispersal would be facilitated. These species then disperse over water to New Guinea and nearby islands and colonize similar marginal habitats there. There they may be extinguished (ending the cycle), or they invade the inner rain forests of these islands. They may also be extinguished at this point, or they adapt evolutionarily to these interior habitats and ultimately diverge so that they are no longer conspecific with Asian populations. Finally, especially on New Guinea, there may be re-adaptation to marginal habitats and the whole process may be repeated. Those species in the process of expansion and before divergence on the islands are called Stage I taxa, while Stage II taxa have diverged on different islands and undergone extinction on smaller islands. Stage III taxa consist of highly fragmented species groups restricted to interior habitats. The point in the dynamic cycle which a taxon has reached is inferred from the stage into which its geographic and habitat distributions fall.

Greenslade (1968a, b, 1969) has shown that this taxon cycle explains well the distributions of land and freshwater birds and several insect species in the Solomon Islands, and Darlington (1970) has interpreted distributions of certain West Indian carabid beetles in a similar fashion. Ricklefs (1970) claims that the taxon cycle on Jamaica is slightly different, in that recent arrivals on the island have widespread habitat distributions as opposed to solely marginal ones, though older populations do tend to become restricted to central habitats as in the above examples.

Finally, a speculative evolutionary interpretation of the equilibrium theory has been suggested by Janzen (1968). In addition to the obvious consideration of host plant individuals as habitat islands in ecological time for individual phytophagous insects, Janzen notes that a host plant species may be considered as an "island" in evolutionary time, with the distance between two such islands for a phytophagous insect species being the evolutionary "distance" between the two plant species. The parallels with the equilibrium model are striking: one might expect that any plant species can support only a given number of phytophagous insect species (the equilibrium), determined by a balance between the number of species utilizing that host for the first time (immigration) and the number of species currently utilizing the host which go extinct. The "immigration" rate should be a function of how "close" taxonomically the host plant is to large groups of other plants, while the extinction rate will be largely dependent on the physical size and population size of the plant.

Paleontology and the Equilibrium Model

Along with most biogeographers, paleontologists have tended to emphasize evolutionary influences on biogeographical events and distributions at the expense of short-term ecological ones. The notion of extinction as an oft-repeated, local event runs counter to the conception in paleontological literature that extinction involves the global termination of a maladapted phylogenetic line, while immigration (as opposed to adaptive radiation) is rarely considered except in the context of cataclysmic events like the establishment or severance of land bridges.

Concurrently, paleontologists have largely, though not exclusively, adopted a qualitative approach to distributional matters, treating a given lineage or group as an exemplar, with scant attention paid to statistical treatment of all available data. Indeed, purely deductive, mathematical models were rarely attempted at all, so that formal numerical predictions were not even generated, much less tested.

The evidence discussed above indicates that at the least the equilibrium model is a useful conceptual framework with which to view species diversity on islands, and a growing body of direct experimental evidence on small insular ecosystems has yet to contradict the tenets of the model. At most, the equilibrium theory may be shown to provide major insights into biogeographic distributions of all types through the formal analogy between a truly oceanic island and any region which is clearly delimited from similar regions.

Research toward this latter end has involved statistical treatment of consequences of the dynamic equilibrium theory (the species-area problem, turnover rates, etc.) rather than direct testing of the notion of a dynamic equilibrium *per se*. Much more work will be required to validate the equilibrium model as a general principle, though the diversity of problems and scales to which it has been applied demonstrates a broad didactic utility, at least.

Paleontologists are, of course, precluded from direct experiments such as those described for small mangrove islands, but can apply their data to the sort of testing discussed in the previous paragraph. Sufficiently numerous contradictions of directly deduced consequences of equilibrium theory would falsify the model as effectively as if the mangrove islands did not return to their original species numbers after fumigation, yet only Webb (1969) has moved in this direction with paleontological data. It should also be emphasized that, although the heralded incompleteness of the fossil record is the paleontologist's most characteristic answer to every anomaly or apparent contradiction, the time span of the fossil record uniquely arms paleontologists with data on turnover rates.

More fundamentally, the apparent success of the equilibrium model in at least several instances ought to convince paleontologists of the utility of abstraction, then mathematical formalization, and finally deduction of testable consequences as a method of generating and answering illuminating questions. Any deductive model represents a compromise among realism, generality, and precision (Levins, 1966); because of the nature of available data the equilibrium theory as applied to oceanic islands in ecological time is very general, moderately realistic, and quite imprecise. Depending on the taxa and regions treated, one might expect paleo-biogeographic models (equilibrium or otherwise) to be a bit less general, about as realistic, and hopefully more precise. But whatever the outcome, the modeling approach should prove worthwhile.

Conclusion

The publication of the equilibrium model has effected a reevaluation of the relative importance of evolutionary vs. ecological events in determining present biogeographic distributions. Primarily, these distributions are now viewed as dependent on ecological phenomena, especially competition and frequent dispersal, with evolution acting gradually to modify the parameters of these ecological phenomena, and thus indirectly the equilibrium distributions themselves.

At least part of the difference in emphasis on ecological vs. evolutionary events can be attributed to differences in objects of study. Highly isolated oceanic islands, a small minority of all islands, formerly received a disproportionate amount of attention because of the rare and unusual species found on them. But it is exactly on such islands that evolution can play such a distinctive, striking role, for the depauperate biotas and infrequent genetic immigration associated with isolation ensure less stringent competition (a short-term ecological event), which allows evolutionary "experiments" to persist. On the majority of islands high immigration rates raise the number of species, and high invasion rates hinder genetic isolation. Ecological interactions are therefore dominant and evolution is no more visible than on continents—deviant phenotypes are more ruthlessly eliminated.

Secondary consequences of this "biogeographic revolution" include the posing of new questions, more quantitative approaches to longstanding questions, and a realization of the similarities between large scale biogeographic theaters and microcosmic ecosystems.

The model is still in its formative stages, however, and a number of theoretical aspects remain to be developed, particularly with respect to the ecological and evolutionary effects of species interactions and the inclusion of information on which species will be involved in various equilibrium phenomena as well as how many species. Additionally, more distributional data on a greater diversity of taxa and ecological data on dispersal are required to validate predictions already made.

Appendix. Table of Symbols

Symbol	Meaning
A	Area of an island.
α	Fitted constant of propagule production. Number of propagules leaving location i per unit time is αw_i^2.
d_i	Distance from recipient island to location i.
D	Distance of an island from the source area for its biota.
e_α	Species extinction rate of extinctions from non-interactive causes for species α. The rate of extinctions per unit time which would be observed if, after every extinction, a propagule of species α colonized the island.
E_s	Extinction rate, in species per unit time, on a given island when S species are present.
G	Constant for rate of colonization.
γ	Number of genera in a source biota.
i_α	Invasion rate for species α, in propagules per unit time.
I_s	Immigration rate, in species per unit time, on a given island when S species are present.
k	Fitted constant of species-area relationship.
λ	Mean distance traveled by propagules of a given species which has an exponential dispersal distribution.
n_i	Number of propagules of a given species arriving on a recipient island from location i.

P_{ij}	Measurement of the jth physical parameter for location i.
P	Number of species in a species pool.
s	Standard deviation of distance traveled by propagules of a given species which has a normal dispersal distribution.
S	Number of species on an island.
$\hat{S}$	Equilibrium number of species for an island.
S_i	Number of species at location i.
$\hat{S}_n$	Non-interactive equilibrium number of species for an island.
$S_{(t)}$	Number of species on an island at time t.
σ	Number of species in a source biota.
t	Time.
$t_{0.90}$	Time required for a sterile island to reach 90% of its equilibrium number of species.
w_i	Width of location i, measured at right angle to line joining i and a recipient island.
$\hat{X}$	Turnover rate, in species per unit time, when an island is at equilibrium.
z	Fitted constant (exponent) of species-area relationship, usually in the vicinity of 0.3.

10

CONCEPTUAL MODELS OF ECOSYSTEM EVOLUTION

James W. Valentine

Editorial introduction. The culmination of biological organization is seen in the ecosystem. In the "evolution" of ecosystems, Valentine carefully distinguished diversity-independent and diversity-dependent factors as causal factors in the observed changes through geologic time. Ultimately, he concludes, "it is changes in stability that cause many of the fundamental changes in the taxonomic composition of the shelf ecosystem."

> The construction of a model that is supposed to represent . . . reality always has to take two conflicting pressures into account; on the one hand, it must be simple enough to make it amenable to . . . stringent methods; on the other hand, it must be sufficiently realistic that the results of studying the model's properties actually lend themselves to the problem at hand. The simplicity of the model—its . . . beauty—is incompatible with its faithfulness to reality. The beautiful are seldom faithful and the faithful are seldom beautiful.
>
> (Alfvén, 1966)

Introduction

An ecosystem is a living association of species populations among which energy flow occurs. However, each ecosystem is an open system which relies upon a fluctuating energy base and which may export and import energy-rich substances at numerous points along the pathways of energy flow within the system. Within any given ecosystem, imbalances may thus exist in the amount of energy in the base and the amount represented by processes within the system and by terminal exports, and also in entropy changes. These factors are balanced by appropriate changes in other energy systems, such as in other ecosystems or even in the sun. Although the energetics of an ecosystem cannot be calculated in the form of a balance sheet of internal energy processes, the flow of energy through an ecosystem is nevertheless the most important single parameter by which to describe its structure and function.

In most ecosystems, energy enters in the form of sunlight and is used to synthesize protoplasm by photosynthesizers, which are therefore regarded as the *primary producers*. In some ecosystems, however, the energy is first introduced in the form of organic or at least organically created compounds which were synthesized elsewhere. In either event a fraction of the energy stored in the producer populations is assimilated by *first-level consumers*, which are either herbivores or detritus-feeders, and a fraction of this assimilated energy passes to *second-level* and eventually to *third-level consumers* and so on up the familiar energy pyramid. The amount of energy available to any level of the pyramid, called a trophic level, depends upon the amount of energy that the populations on preceding levels have stored.

In size, ecosystems may vary from a few species populations living in a microhabitat, such as on the stipe of an alga, to the entire biosphere. At the biosphere level the ecosystem is composed of all living organisms, energy input is based upon the solar radiation falling on the earth's surface, and nutrient cycles involve the oceanic and atmospheric circulation and also geochemical cycles. For our purposes we shall consider ecosystems on the level of the community; they are composed of living populations closely associated in space and time within a certain discrete range of habitat conditions, such as a subtidal mud flat or a sandy beach.

It is possible to imagine a wide variety of energy flow patterns within ecosystems, giving rise to a variety of ecosystem structures. Each trophic level may contain few or many different populations, which may be more or less efficient, and which may utilize few or many of the populations in the preceding level as prey, and each may be preyed upon in turn by few or many of the populations in the succeeding level. Furthermore, some populations may draw their energy supplies from more than one trophic level, and therefore the energy they represent has to be divided among different levels in appropriate proportions. If we examine living ecosystems in the present oceans, we find that there is actually a great variety of structural types present, and they appear to vary in a systematic manner with variations in those physical environmental parameters that most affect the energy base. Thus the structure of ecosystems is responsive to the environment, and therefore the energetic contexts within which the populations of any ecosystem exist, and the biological context as reflected in the numbers of and sorts of species with which each population must contend, all vary environmentally. These are fundamental components in the adaptive spectrum, and the significance of ecosystem structure in determining the evolutionary pathways of lineages must be very great indeed.

An ancient ecosystem is preserved as an assemblage of fossils that represent only a fraction of the living association that once functioned in a trophic pyramid. Furthermore the fossils are normally mixtures of different generations and often include species that lived in the ecosystem at separate times or that have been intermixed from other ecosystems. To reconstruct the energetics of such ancient systems would appear to be an impossible task given the imperfections of the data, and in a way it is. On the other hand, it may be possible to construct *models* of ancient ecosystems and of their evolution, models that have important explanatory powers. In fact a similar approach is employed with living ecosystems. The details of productivity, standing biomass, food sources, predator-prey interactions, scavenging systems, and so forth are not understood for all populations within any living ecosystem and are known in a general way for only a precious few. The descriptions of living ecosystems, therefore, are models themselves, which depict only the general patterns of energy flow and population interactions. Commonly they are based upon only a single taxonomic group—birds or butterflies, for example, are studied closely, and then models of ecosystem structure, function, and regulation are erected. To develop models of ancient ecosystems is therefore not much more difficult, for we may employ the best-preserved fractions. Modeling would appear to be a promising approach to the reconstruction of the evolution of energetic patterns. We need not despair because we do not have every last member of the original community as fossils, or even because the soft-bodied fractions are essentially unknown. If we are careful, and base our models on the best known and most representative groups of the skeletonized fraction, we may still be able to form satisfying scientific explanations that have a high probability of being correct.

Sources for Models of Ecosystems

As the data on ancient ecosystems are so fragmentary it is usually necessary or at least convenient to draw ideas as to their functions from a variety of sources outside the fossil remains themselves. Rudwick (1964b) has suggested sources for functional models of the structures of individual organisms that may be modified to suit functional models of ecosystems. A logical source of ideas would be in living ecosystems that closely resemble the fossil ones, insofar as the comparison is possible. Failing to find such a particular living ecosystem, one may use any living system that shares some special attribute with the fossil ones, and perhaps a model may be built up from a number of such sources. Or physical systems that lack biological components suggest ideas to aid in modeling ancient ecosystems. An example of such a physical system is an electric circuit, which can be used as an analogue of energy paths in ecosystems (Odum, 1960). Analogies can also be suggested with artificial systems such as economic systems (Simon, 1970; Tullock, 1971). And finally, ideas may be derived from sheer invention, without any obvious source in the real world.

A model built up from such sources should obviously be subjected to some sort of test. Rudwick (1961, 1964b) has suggested that models of the structures of individual organisms should be conceptualized so that they would operate as efficiently as possible, given the nature of the materials. This efficient model is then compared with the structure to be interpreted. A close fit would indicate that the model was possibly correct in essential features. This is a form of parsimony, an attempt to limit the possible number of explanations to the simpler ones required by the data. The same sort of approach may be employed in modeling ecosystem states and evolutionary pathways. The simplest model that explains the facts is to be preferred, and this implies the most efficient model insofar as functional attributes are concerned, considering the nature of the environmental regime. Let us examine models of the structure of a variety of modern marine ecosystems, in the hope that they may help in erecting functional models of ancient ecosystems. As most fossil assemblages lived in shallow depths, we shall restrict our models chiefly to ecosystems of the continental shelves.

Models of Diversity Gradients and Diversity Trends

One of the major aspects of ecosystem structure is the number of species present, called the species diversity. Diversity is clearly related to the way that energy is partitioned and to the pattern of its flow within ecosystems. Most populations that live within ecosystems of high diversity have quite a different biological environment from those that live within ecosystems of low diversity, and it is to be expected that there are systematic, qualitative differences between populations in these different situations. It is worth while therefore to examine the causes of diversity differences in living ecosystems. It turns out that there is no single, well verified theory of diversity regulation, although there are a number of models (chiefly based on diversity patterns within certain taxonomic groups). A large number of workers have contributed to the development of these models, including in more recent years Fischer (1960), Hessler and Sanders (1967), Klopfer and MacArthur (1960), MacArthur (1955, 1957), MacArthur and Wilson (1963), Pianka (1966), Sanders (1968, 1969), Valentine (1969a, 1971a, 1971b), and Valentine and Moores (1970). It is not necessary to review here each contribution to the models, but it is of interest to review a few of the important hypotheses that have sought to explain aspects of diversity.

Temperature is the most obvious environmental parameter that correlates to some extent with the latitudinal diversity gradient; it is warm near the equator where diversity is high and cool near the poles where it is low. It therefore seems possible at first thought that the temperature gradient is somehow responsible for the diversity gradient, although the mechanisms of this diversity control are obscure. Temperature is certainly a major factor in limiting the geographic ranges of species of marine invertebrates (Hutchins, 1947; Gunter, 1957; Kinne, 1963), and this fact seems to lend credence to the hypothesis of a temperature control of diversity. There have been suggestions that higher temperatures are more optimal for protoplasm or cause higher rates of evolution, and therefore more kinds of organisms live in the tropics. However, the obligatory correlation of low temperature with low diversity has now been thoroughly disproven by the discovery that the deep-sea communities are highly diverse relative to comparable shelf communities (Hessler and Sanders, 1967; Sanders, 1968), yet the deep-sea communities live at temperatures of over 25°C cooler than tropical muddy-bottom shelf communities.

Another hypothesis involving temperature suggests that it is not the temperature level but the temperature fluctuations that control diversity, and that thermally stable communities (tropical shelves, deep-sea) may be highly diverse, while thermally unstable communities such as are found on high-latitude shelves must be less diverse. From low to middle latitudes, decreasing diversity does correlate with increasing seasonal temperature differences. However, in proceeding to very high latitudes the seasonal differences begin to decrease again, as summer water temperatures become cooler and cooler while winter water temperatures cannot decrease much since they are near the freezing point. Thus waters in very high latitudes are rather stable in temperature. Yet diversity continues to decrease in higher latitudes even though temperature ranges are decreasing also. The correlation does not hold and the temperature stability hypothesis is not supported.

In neither of these cases is there a strong biological rationale for the temperature control of diversity. Usually it was merely observed that few species lived where temperatures were low or where temperature ranges were high, and then inferred that only a few species could tolerate one or the other of these conditions and therefore diversity was low where they prevailed. What is lacking in these hypotheses is a reason that prevents natural selection from developing a large number of species that are adapted to the appropriate temperature level of the regime, and filling those environments that have low or fluctuating temperatures just as full of species as those that have high or stable temperatures. In other words, temperature is a *diversity-independent* factor. It may operate to restrict the access of some species to certain environments, but it does so without regard to the number of species already living in these environments. The fact that an environment is hot, or cold, or thermally fluctuating, would not of itself prevent increasingly large numbers of species from living there, if natural selection were to develop more lineages with appropriate tolerances. At least, no compelling reason for such prevention has ever been advanced.

Another common hypothesis to explain diversity patterns is called the *time* hypothesis, which is frequently applied to the very high latitude ecosystems which are of low diversity. The argument usually runs something like this: The appearance of very cool poles, with high-latitude continental ice caps or sheets and widespread winter sea-ice, is a relatively recent phenomenon, geologically speaking. Natural selection has not had *time* to develop a diverse biota adapted to these conditions, although sooner or later it will, and then diversity in those conditions will be much higher than at present, perhaps even as high as in the tropics. Whether or not the highest latitude communities have indeed reached their full potential diversity (assuming that there is such a thing) is certainly not known, but there are reasons to doubt that the argument of *time* is sufficient to explain their present states. For example, the diversity gradient is present within the tropic, subtropic, and temperate zones, and yet many of these climatic zones must date at least from the late Paleozoic and perhaps from the Precambrian. It is not obvious that the subtropical climates are younger than the tropical, or the warm temperate climates younger than the subtropical. Yet the diversity gradient appears within and across these climatic regions. There would seem to be no special reason to plead for time as a factor in the continuation of the same gradient into higher latitudes.

Finally, a common hypothesis is that diversity control cannot be reduced to a single factor but is owing to general environmental stability, and that fluctuation in any or all parameters is inimical to high species diversities, which can be developed only in stable environments. This hypothesis has been especially championed in recent years by Sanders (1968, 1969) and Bretsky and Lorenz (1970), who present a model of the different genetic strategies that may underpin the adaptive strategies found in stable versus fluctuating environments (see also Levins, 1968). In order that this hypothesis not be falsified, it is necessary first of all to show that environmental fluctuations are indeed greater in high latitudes where diversity falls.

Although temperature variations are not greatest in higher latitudes, as we have seen, there are certain other environmental variables that reach their greatest fluctuations there, or at least their greatest on a regional scale in the oceans. Salinity variations, for example, arise seasonally from melting and freezing ice. This factor affects a great many water properties—the colligative properties, for example—and thus may have widespread ecological consequences. And another factor that must in fact have its greatest variability in the highest latitudes is solar energy. This variable is of special interest, for it varies from near stability at the equator, where days are about equally long all year, to fluctuations at the poles that are so great that days are seasonal. General stability appears to correlate with longitudinal variations in diversity also (Sanders, 1969). Shelves that have highly variable "continental" climates appear to support fewer species than shelves having less continental, more equable "maritime" climates, even at identical latitudes where solar radiation is of course essentially the same. Thus the general stability hypothesis appears to be capable of explaining marine diversity patterns. However, the question remains, in G. G. Simpson's phrase, how come?

We have seen that diversity-independent factors such as the parameters associated with the temperature regime cannot by themselves account for diversity controls. Therefore, unless it can be shown that they are effective in some combination or with some other factors, they cannot be considered as functioning parts of the general environment insofar as diversity regulation is concerned. Now, diversity-independent factors could limit diversity if they occurred in an unpredictable fashion and were rather severe. Thus a regional change in, say, temperature regime might cause local extinctions of some species, and while they were being replaced by species with appropriate adaptations to the new temperature regime, another temperature change or a change in some other parameter would exclude still other species, so that the biota would never be able, owing to irregular and unpredictable environmental changes, to contain many species with appropriate adaptations (Fischer, 1960). This is a conceivable mechanism or model of diversity control by factors that are in fact diversity-independent. Indeed, in an earlier contribution in this volume Johnson has indicated how communities and even portions of communities may be affected by such factors. Nevertheless, it is doubtful whether they can account for the major diversity trends in the ocean at present. Species in low and intermediate latitudes, where a latitudinal diversity gradient is well displayed, do not appear to be subjected to these kinds of environmental changes. Species numbers do not appear to be regulated by a process of local extinction due to series of unpredictable environmental fluctuations.

In short, it seems that if environmental variability controls species diversity over a long period of time, there must be some factor present that is *diversity-dependent*, that is sensitive to the number of species that are present in an ecosystem, and that permits fewer additional species to enter as the number present rises to some maximum level beyond which no more can coexist for long. Two environmental factors that most closely approximate this requirement appear to be (1) habitat space and (2) trophic resources (that is, sunlight and nutrients for primary producers at the base of the trophic pyramid and organisms or organic detritus as food at each succeeding level). To understand why trophic resources may be considered as diversity-dependent, it is necessary to examine still other models, models of adaptive strategies.

Models of Adaptive Strategies for Trophic Resources

Ecosystems are composed of populations that are adapted to the ambient environmental regime. Thus if the regime favors populations that are large or small, omnivorous or trophic specialists, ubiquitous or restricted in habitat, the ecosystem will contain that sort of species. Clearly, the structure of the ecosystem will grow out of and depend upon the adaptations of its component populations. When we model an ecosystem, the structure that we build into the model implies a certain set of adaptive strategies among the populations that make it up. Thus it behooves us to examine the relationship between environmental conditions and adaptive strategies.

A large number of investigators have considered the effects of trophic resource utilization, both implicitly and explicitly. In addition to several that were previously listed as contributions to diversity theory, the papers of Connell and Orias (1964), Lewontin (1957, 1958), and Margalef (1968) should be mentioned. If the resource regimes are important, it is because they require special adaptive strategies of the populations they support. The term "adaptive strategy" is not meant to be teleological, to suggest a purposeful creation of a plan of adaptation by an evolving lineage. It merely denotes the results of natural selection for adaptation to an environmental pattern, such a pattern as is imposed by spatial heterogeneity or temporal fluctuations in some factor (see Levins, 1968; Bretsky and Lorenz, 1970). Here we are chiefly concerned with temporal patterns in resource supply.

In environments wherein trophic resources fluctuate greatly, populations must be prepared to survive during the lean periods when resource levels are low, perhaps at zero. Mortality might be expected to be high during periods of little or no energy supplies. Populations may weather these periods by simple fasting, by lowering their metabolic rates as in hibernation, by reproducing and lasting out the inclement period as dormant zygotes, by emigrating, or by other means. For those populations that remain to face the inclement periods, it seems advantageous if they are large in numbers, for then they have the best chance that some of their numbers will survive. There are several obvious components of an adaptive strategy towards this end.

One is to have a high reproductive potential, so that the population may be increased rapidly whenever conditions improve, in order to be prepared as soon as possible for any subsequent deterioration. Another is to be able to eat as wide a variety of prey as possible, in order that the failure of one or of a few particular populations will not affect the energy supply disastrously. Whatever species happen to flourish on the next lower trophic level, they are available as resources to trophic generalists. Still another adaptation is to be able to live in as large a variety of habitats as possible. If a population is not restricted to a narrow range of habitat conditions, so that it soon fills up the available habitat space, it may spread widely and maintain a large size, and thus be better prepared for inclement periods. Furthermore, the more widely it is spread in the more diverse types of habitats, the greater its chances of being represented in some area that happens to be less poorly off than average, in the event of an unusually inclement season, and thus the chances of its survival are improved. All of these components add up to an adaptive strategy model of a generalized, flexible population, one that is adapted to a broad range of environmental conditions. Such populations are very opportunistic, and MacArthur and Wilson (1967) consider species with these attributes to be good colonizing species (see also Stebbins, 1950).

In environments with relatively stable trophic resource supplies, on the other hand, the populations are not required to undergo periods of fasting, by definition. As their food supply is assured, the necessity for maintaining large populations has vanished, insofar as trophic resource supplies are concerned. Therefore a low reproductive potential may do very nicely. (Of course there may be other reasons why a high reproductive potential is evolved—to offset predation, to aid in dispersal, and so on.) Similarly, it would be possible to eat only a single (or a few) prey species, since they are going to be in steady supply. And possession of only a very narrow range of habitats is not in this event a prelude to extinction, since a low-density, patchy dispersion pattern is not particularly disadvantageous. It may even aid in predator avoidance. At very low population densities it may be necessary to develop some special reproductive strategy to insure fertilization, but that is another story. For that matter, specializations for predator avoidance, reproduction, or other functions may result in lower resource demands by a species and therefore permit diversity to rise by freeing resources for other species.

Models of Ecosystems

If these different strategies are actually followed by organisms under different trophic resource regimes, then ecosystems in environments wherein resources fluctuate should be composed of relatively broad-niched species.

These would be rather inefficient ecosystems, for they would have a boom-and-bust economy, with biomass springing up at high rates during high resource periods and crashing to low levels during low resource periods. The fluctuations would be greater at higher trophic levels, so that only a few levels could be supported in a trophic pyramid. Indeed, only a relatively few of these generalized, inefficient populations could be accommodated within a given range of habitat conditions. If a new population immigrated into such an ecosystem it would require a share (and a large share at that) of the available resources as it would have to support a large population in order to survive. But this means that some other population or populations must have reduced energy sources, losing some of their share, and this would endanger them, considering the requirements for success in living on fluctuating trophic resources. Sooner or later a particularly bad period would cause the extinction of one of these populations within the ecosystem, and then the remaining species could return to their former shares of the trophic resources and diversity would be back at its former level. In resource-stable environments, ecosystems would have higher efficiencies with less "wastage" of biomass and with stability continuing to higher trophic levels, so that relatively large numbers of specialized, efficient populations could be accommodated within a given range of habitat conditions. If a new population immigrates into such an ecosystem and appropriates a share of the available resources, there is a good chance that it can be accommodated indefinitely by a rationing out of its share among many species. Although this will lower their standing population sizes on the average, this is not necessarily a critical factor in their survival. Therefore diversity in trophically stable environments should be higher than in trophically unstable ones. According to these models of adaptive strategy, the relative stability of trophic resources represents a diversity-dependent factor.

Besides their stability, another aspect of trophic resources that may prove to be important in diversity regulation is simply their level. Is a *high food supply* conducive to high diversity or low, or is it neutral? According to the models of adaptive strategy that we have discussed, it is likely that high trophic resources tend to favor low diversity. This is because when food is in large supply, populations with high reproductive potentials and generalized habits may multiply and usurp the habitats that specialists would normally occupy at low resource levels. That is, so long as resources are low, the specialists are clearly superior (in theory), for they are extremely well adapted and adept at utilizing their own special food sources. However, when food is not particularly limiting, this advantage is lessened, since even a generalist may feed easily, and competition for habitat space or some other environmental factor may become severe. An energy increase in a stable ecosystem does not permit more populations to appear, but merely permits those populations with the highest reproductive potential to multiply, in which case they probably will tend to overwhelm some of their more specialized neighbors (Margalef, 1968; Valentine, 1971a).

Ecosystems in fluctuating environments, then, may have a trophic structure similar to that in *figure 10–1A*. There are few different species, but most feed on a large number of populations, so that the energy paths resemble a web.

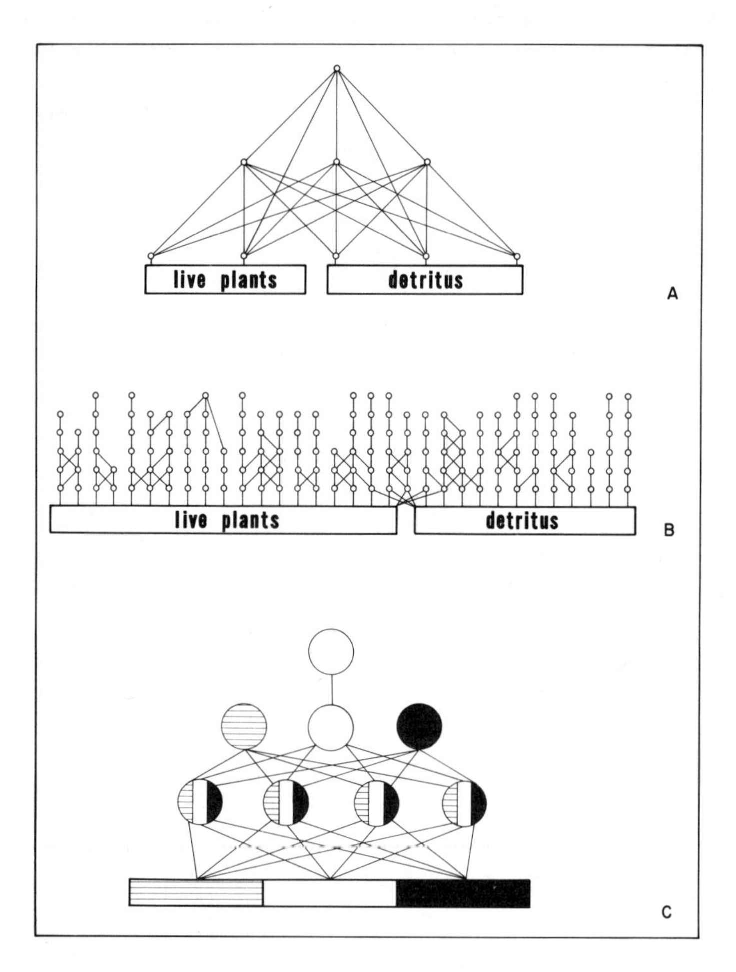

Figure 10–1: Some diagrammatically exaggerated trophic structures; consumer populations are indicated by circles. (A), an ecosystem with a fluctuating trophic resource base. (B), an ecosystem with a stable trophic resource base in which resource specialization is the dominant adapative strategy. Energy is narrowly channeled along chains of specialized consumers. (C), an ecosystem with a stable trophic resource base in which habitat specialization is a dominant adaptive strategy. Energy is channeled within different habitats, three of which are indicated by the different shadings.

In such an unstable environment, a good strategy is to feed upon the most stable resource possible, and this is commonly detritus. Live phytoplankton may be present in abundance only for a few months or less, but detritus is more likely to be available the year round. Furthermore, detritus-feeding is surely about as generalized a trophic condition as is possible, although, to be sure, it is possible to specialize in one sort of detritus. Nearly every kind of organism contributes to the detritus. Therefore, in highly fluctuating environments, detritus-feeders and perhaps scavengers should make up a large proportion of the feeding types. This appears to be the case, both in the Arctic (Valentine, unpublished) and Antarctic (Arnaud, 1970). It follows that ecosystems of low diversity should commonly contain a relatively high proportion of detritus-feeders and scavengers, and that much energy flows through such populations.

Ecosystems in stable environments have a trophic structure more as depicted in *figures 10-1B* and *10-1C*, if the model is correct. There are many different species, and most either feed on only a few other populations, or live in only a few special habitats, or both. As resources are stable it is possible to exploit living populations directly rather than wait for them to generate detritus. In fact it is a very good strategy to do this before some other population does it. A special competence at securing a particular living prey would be good insurance against competition. If an ecosystem in a stable environment consists chiefly of trophic specialists, energy tends to flow along chains of these specialists, partitioning the primary production and canalizing the energy to some top predator quite efficiently in relative terms (*figure 10-1B*). If an ecosystem consists chiefly of habitat specialists, then energy is canalized, not through populations, but through local habitats in an areal patchwork (*figure 10-1C*). This means, among other things, that many different species will be encountered in different localities during a traverse of the community. Tropical shelf communities seem to contain both habitat and trophic specialists, so that their trophic structures are characterized by a combination of the properties sketched in *figures 10-1B* and *10-1C*. If most populations in a stable environment were regulated by other specializations, such as reproductive ones, then the trophic structures of their communities would have still different properties.

The ecosystem structures that are depicted in *figure 10–1* have been derived chiefly by deduction from the assumptions of the previous models of diversity control and of the related models of adaptive strategies. Although they are satisfying in some ways, it is prudent to test them by comparison with living ecosystems before attempting to employ them as an aid in interpreting fossil ecosystems. One difficulty is that, since few living ecosystems have been thoroughly investigated, we are for the most part comparing our deductive model with other deductive models that are also based on fragmentary data and in fact that share some of the same basic data and assumptions. Such a comparison has a large component of circularity. We clearly need to go to the ocean and study energy flow in many living marine ecosystems; data on the tropics would be particularly valuable. Such studies as are available tend to support the models in a general way. One somewhat independent test is in palagic ecosystems, which have been ignored in constructing the model; in these, high-resource bases in unstable waters such as upwelling regions do indeed support ecosystems of low diversity and few trophic levels, while stable and low-resource bases support ecosystems of high diversity and more trophic levels (Ryther, 1969). Freshwater ecosystems that have been carefully studied in the African Great Lakes exhibit similar relations (Fryer and Iles, 1969).

There is, however, one obvious aspect of living marine benthic ecosystems that appears to place a difficulty in the way of our model. This is that some specialized populations do live in fluctuating environments, and some generalized ones live in highly stable environments. If one specialist can live in the Arctic, say, why then perhaps many others eventually will, and the ecosystems there will become diverse after all. A possible answer to this dilemma is illustrated in *figure 10-2.* Resources in fluctuating ecosystems are not perfectly unstable, nor are resources in any real ecosystem perfectly stable. Therefore, in fluctuating environments, a certain fraction of the trophic

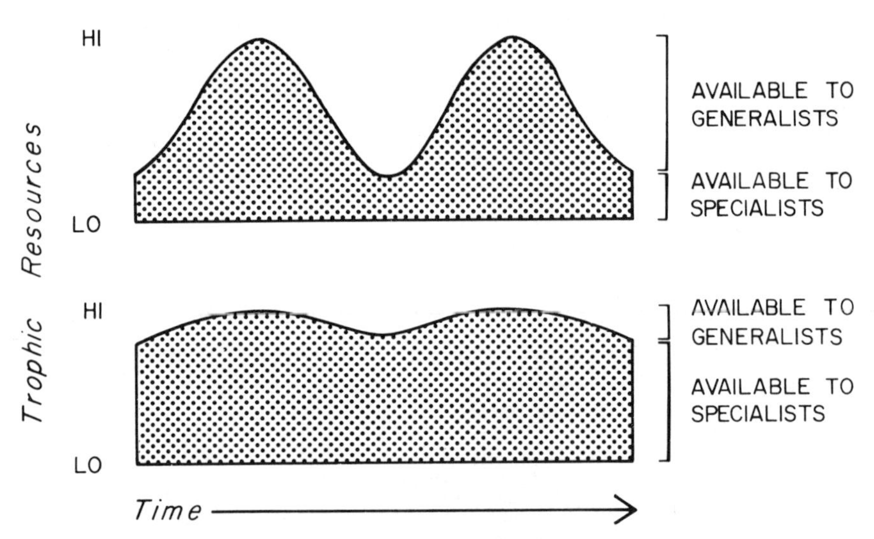

Figure 10–2: A simplified model of the changing proportions of trophic strategies that may be found in less stable vs. more stable resource regimes. Even when most of the resources are fluctuating, some small stable fraction should be available for trophic specialists, while even in rather stable regimes some small fraction of the resources fluctuates and can be utilized only by opportunistic species.

resources remains stable and is available for exploitation by specialists. In stable environments, on the other hand, there is some fraction of the trophic resources that fluctuates. This fraction is unavailable to those highly specialized populations that have based their adaptive strategies on the assumption that resources are stable. It may, however, be utilized by trophic generalists with flexible strategies, and therefore such species can exist even within rather stable tropical ecosystems. For a trophic generalist the diverse ecosystems of the tropics must present a virtual smorgasbord of gourmet delights. These generalists are the opportunistic species of any ecosystem, which appear when certain conditions are met, including an appropriate surplus of energy, and disappear when conditions change. Levinton (1970) has discussed the impact of such species on the fossil record. They may appear in vast numbers for a season or two and then disappear, leaving behind abundant skeletal debris that suggests to the paleontologist that they were a dominant element in the ecosystem structure when in fact they are more like an unstable fringe. However, they exist and are important, and if they can be identified as opportunists they can add much to our ability to interpret the past.

And finally, there has long been a dispute as to whether or not it is possible for "vacant niches" to be present within ecosystems, and the present model bears upon this problem. The dispute has sometimes involved a semantic question but clearly involves substantive problems as well. To state it in the baldest terms, the question is whether there may exist, within the economy of an ecosystem that is more or less in equilibrium, a place that is unfilled, a potential function that is not realized, so that if members of an appropriate species found their way into this ecosystem they could find easy acceptance without significant repercussions in any other part of the system—they could fill a "vacant niche." According to the present model this is probably not possible, depending upon the emphasis given to the term "equilibrium." This is an important point because if "vacant niches" could exist at equilibrium, then diversity fluctuations could occur merely by the creation or filling of these, and need not at all reflect the operation of an ecological regulator.

To make the model work at all it is necessary to assume that organisms possess some sort of potential that usually obliges them to fill as much of the environment as they are permitted. Such an assumption for the level of the ecosystem is closely analogous to and depends upon the assumption on the level of the population that there is a certain "biotic potential" that would lead to vast increases in population size were it not for limiting factors in the environment. This assumption worked well in establishing the theory of natural selection, so we shall employ it as a premise in our model. Therefore, we assume that in a given environment, diversity will tend to rise until checked by some limiting factors. If the factors are diversity-independent, a population that is extinguished does indeed leave behind it a potential opening for another population—an unused energy base, a vacated habitat, and so on. This may be filled by a species that is new to the ecosystem. Thus the "vacant niche" exists, but only for the length of time that it takes to generate a replacement, and in terms of geological time this is probably a very short period. If the factor that limits diversity is diversity-dependent, then an opening does not exist, even after diversity is lowered from a previously higher level. However it may *appear* that an opening exists, if a parameter that is not limiting is observed by an investigator.

For example, the two major diversity-dependent factors are trophic resources and habitat space. If the former is limiting in some relation, then there may be habitats that are unoccupied and that are, indeed, filled with species in similar ecosystems at other places or in other times. This gives the impression that there is a "vacant niche." However, there is in fact no "room" in the ecosystem for more species, for considering the instability of the resource regime the available trophic resources are partitioned among as many populations as they can be stretched to cover, and adding populations will merely cause some of the previous occupant populations to become extinct sooner or later. If on the other hand the habitat space is limiting, there will be an excess of trophic resources and it will appear as if food were being wasted, although in fact the available habitat space is full. Such situations would be most likely to occur in spatially homogeneous environments, such as vast mud flats.

Although we cannot check our model in detail against many living ecosystems, as they are not well known, it is still possible to check the predictions of diversity that arise from the model against the major diversity trends and patterns which are known for the living benthic realm. We have already noted that there is a good correlation between latitudinal trends in diversity and stability of solar radiation. At any given latitude, according to our model, diversity should be nearly constant when similar ranges of habitats are compared, insofar as solar radiation is concerned. Any diversity trends that do exist longitudinally must be due to some other factors that serve to perturb the trophic resources. In the oceans it is common for primary productivity to be limited by nutrient supply. Therefore fluctuations in nutrient supply from place to place longitudinally could act much as changes in the stability of solar radiation act latitudinally.

Nutrients in waters over continental shelves commonly display a seasonal pattern owing to seasonal perturbations of the water column or to seasonal changes in water types. For example, along coasts which experience offshore winds in winter and onshore winds in summer, there may be a switch in the hydrography from winter upwelling of nutrient-rich water to a summer supply of nutrient-poor water that is blown onshore from the open ocean. Thus whatever seasonality is provided by solar radiation is further complicated by a seasonality of nutrients, and the resultant pattern of productivity can involve irregular but periodic fluctuations. In this event the adaptive strategies favored by the resource regime should result in relatively low species diversity within ecosystems. On the other hand, shelves may be bathed by a stable current season after season with little change, which should favor adaptive strategies that lead to high species diversity within ecosystems. The strongest perturbations of the water column that are most likely to lead to nutrient fluctuations are to be expected along the coasts of the larger continents or along the shores of the smaller oceans—wherever continentality is high. Water columns should be more stable in general along large oceans or the shores of small continents or islands, where maritime conditions prevail. Therefore, at any given latitude, we would expect diversity to be greatest on shelves bordering the smaller continents in larger oceans, all else being equal.

Judging from the numbers of species of large phyla (such as the Mollusca) that are described from present provinces, longitudinal diversity trends on the continental shelves today agree fairly well with this expectation (Valentine, 1971b). The highest diversities in the shelf realm are recorded in the Indo-Pacific along the chains of island arcs, the shelves of small continents and the drowned shelves of larger continents that stretch from Indonesia north to the Philippines and east to the Solomons and even to Samoa. Diversity is lower in the waters surrounding isolated islands and island chains even in low latitudes, but this is most likely owing to the difficulty of dispersal to these regions and to their restricted areas and habitat ranges. At any rate, this high-diversity region is composed of smaller lands situated in the largest present ocean. The next higher diversity appears to be off northwestern South America, where a large continent faces a large ocean. And the least diversity is recorded in the equatorial Atlantic, where large continents face a relatively small ocean.

Similar correlations and trends in longitudinal diversity appear to hold at other latitudes. For example, it is possible to compare latitudes in the low 70's between shelves bordering the Arctic Ocean and the Antarctic shelf. The Arctic Ocean is small and bounded by large continental masses, while the Antarctic is a small continent bounded by vast oceanic expanses along much of its shores. Both have low diversities, as would be expected in such high latitudes, but the diversity within communities around Antarctica appears to be higher than in the Arctic, although data are scarce. The ice cover in the Arctic operates to extend the period of low photosynthetic activity and thus to enhance the trophic resource fluctuations normal to those high latitudes (see Dunbar, 1968).

Thus we have identified a factor—the trophic resource regime—that correlates with diversity trends in the present oceans and that has a biological explanation or rationale which arises from considerations of the evolution of adaptive strategies. This is a diversity-dependent factor and may be hypothesized to account for the correlation of diversity patterns with patterns of general environmental stability. Under this hypothesis the resource regime is the factor that actively regulates major trends in diversity, on the continental shelves. This is not to say that other factors such as temperature or oxygen tension cannot control local or even regional diversity at times; diversity-independent controls certainly operate and a region that was, say, chiefly depleted of its oxygen supply would obviously experience a fall in diversity. The model does hypothesize that trophic resource regimes account for a major portion of the general pattern of diversity within ecosystems, and a model of the expected marine diversity pattern in a world with the present continental configuration fits the observed pattern quite closely.

Perhaps the most striking apparent exception is the reef ecosystem, where energy flow is exceptionally high (Sargent and Austin, 1954; Odum and Odum, 1955) and yet diversity is high also. This situation may be accommodated in the model by postulating that the resource stability is so high that effects of the high energy flow are thoroughly masked in the high-diversity levels that are based upon the stability of resources. However, other hypotheses can be offered that point toward possible solutions to this anomaly. One of them involves the use to which the energy is put in this unusual ecosystem. The source of the unusually high primary productivity in reefs is not precisely known. It is certainly not the plankton (Kohn and Helfrich, 1956). Although benthic reef algae are fairly productive (Bakus, 1967; Marsh, 1970), a major fraction of reef productivity must be attributed to the zooxanthellae, protistan autotrophs that live in the tissues of many marine animals and that contribute trophic energy to their hosts (Muscatine and Cernichiari, 1969). Cowen (1970) has proposed an adaptive model to explain the special advantages of zooxanthellae. According to this model, the energy derived from the zooxanthellae permits rapid growth, attainment of large body size, and deposition of a large skeleton. These features are especially useful to reef organisms, for the survival of the physical structure of the reef itself depends in great measure upon the ability of its component populations to counter the destructive effects of wave attack and of biological erosion in order to maintain a wave-resistant framework. Therefore we may postulate that the energy supplement represented by zooxanthellae is employed in maintaining the physical reef structure and not the trophic structure of the reef communities. Indeed, these tissue-dwelling protists are not available to support herbivore populations in the usual way. They must be regarded as contributing to general resource stability, however—harboring an energy source in their own tissues must provide the hosts with an unusually reliable fraction of their energy uptake.

Models in Ecosystem Evolution

Ecosystems can change in a bewildering number of ways, some of which involve organic evolution among their component populations, and some of which do not. To bring a certain amount of order into our discussion it is convenient to class the changes in ecosystems into three divisions: (1) changes that alter the quality of the populations present; (2) changes that alter the proportionate representation of the populations; and (3) changes that alter the diversity of the populations (Valentine, 1968). As we are dealing with functional units, we are especially interested in the functional aspects and consequences of these sorts of ecosystem changes.

Alterations in the functional quality of populations as expressed within an ecosystem can be due to changes in gene frequency within the gene pool that result in changes in the range or frequency distribution of functions within the population. Such changes are usually owing to natural selection but can also be due to genetic drift. On the other hand, population quality can be altered without corresponding genetic changes but merely as a response to an altered environment that evokes functional responses which, although not required by the previous environment, nevertheless were within the potential of the gene pool to realize in appropriate circumstances. In other words, the qualitative changes may be due either to genetic changes or to strictly phenotypic changes that were evoked by environmental changes. Finally, the immigration of species into ecosystems in which they have not been previously represented, or the elimination of some species from an ecosystem, will change the quality of that ecosystem.

Alterations in the proportionate representation of populations may result from changes in population quality, which may permit a population to enlarge or shrink its size relative to the community as a whole; other changes in population proportions may result from changes in the environment that favor populations differentially. These may be changes in habitat proportions or in food supplies (which must be based in turn on some other change) or in density-independent factors such as temperature or salinity.

Alterations in the diversity of populations—in the number of different species that are represented by populations in an ecosystem—are in a way a special case of population proportions, for diversity changes involve increasing the representation of some populations from zero to some fraction, or decreasing the proportion from some fraction to zero. And as alterations in diversity involve the immigration or local extinction of populations, they also result in qualitative changes in the ecosystem.

Let us now imagine the evolution of an ecosystem entailing changes in all of the parameters that we have discussed. If we begin with an ecosystem at equilibrium, we must perturb the environmental regime in order to cause significant changes. The environment may already have a fluctuating regime, and if so the resident populations will be adapted to tolerate the fluctuations and to survive as populations; the normal fluctuations do not put selective pressure upon the resident populations because their adaptive strategies are appropriate to the ambient regime. When the regime is altered, even if this involves a decrease in fluctuations, a different mixture of adaptive strategies becomes appropriate. Some of the populations may continue as before, exploiting any fraction of their potential resources that fluctuates about the same as before. Other populations, faced with rising stability of their resource bases, may evolve so as to lower their reproductive potential and thus conserve some of their energy, thereby increasing their ecological efficiency. This change might entail corresponding coadaptive changes in other ecological functions as well, leading perhaps to increasing habitat and/or prey specialization. As the resources stabilized, and as the functional ranges of specializing lineages shrank, the potential diversity of the ecosystem would increase, so that chance immigrants would now have increased opportunities of achieving more or less permanent colonization, assuming that they were appropriately adapted. The ecosystem is changing in the proportion, quality, and diversity of its component populations; its structure is changing, including the pattern of energy partitioning and flow; and the functional range of the whole ecosystem is changing in response to the new opportunities and requirements of the environment—in short, the ecosystem is evolving.

To trace the expected evolutionary pathways of ecosystems from the principles and processes involved in our model is now a trivial exercise. It is easy to imagine the changes that would accompany variations in the level or the stability of trophic resources or of other diversity-dependent factors, as well as those that would accompany changes in the regimes of diversity-independent factors.

Although we are chiefly concerned with building a conceptual model, and have now put together a provisional one, it is certainly necessary to test it by examining the fossil record. All too few ancient ecosystems have been described, even to the extent of having an ecological description of a restricted group of taxa from some sequence of ecosystems that could serve as the basis of a model. Perhaps the best way to make a preliminary test of the model (that is, to try to falsify it) is to examine the major patterns of diversification, extinction, and diversity levels throughout the Phanerozoic, to see if they can be interpreted in a manner that is consistent with the model. These patterns have been depicted and discussed by Newell (1956, 1967) and Valentine (1969b), among others; the major trends for well-skeletonized benthic shelf invertebrates are shown in *figure 10–3*.

If it is assumed that the major fluctuations in the taxonomic diversity of the fossils are due to ecological diversity regulation, then the model can be used to interpret the changes in diversity and to predict the conditions that should accompany them. For example, the two major diversity-dependent factors would produce the following effects. First, diversity limitations that are imposed by rising trophic resource fluctuations should be characterized by a lowering of diversity that is not quickly made up by rediversification, but that remains low for so long as the limitations endure. The fauna associated

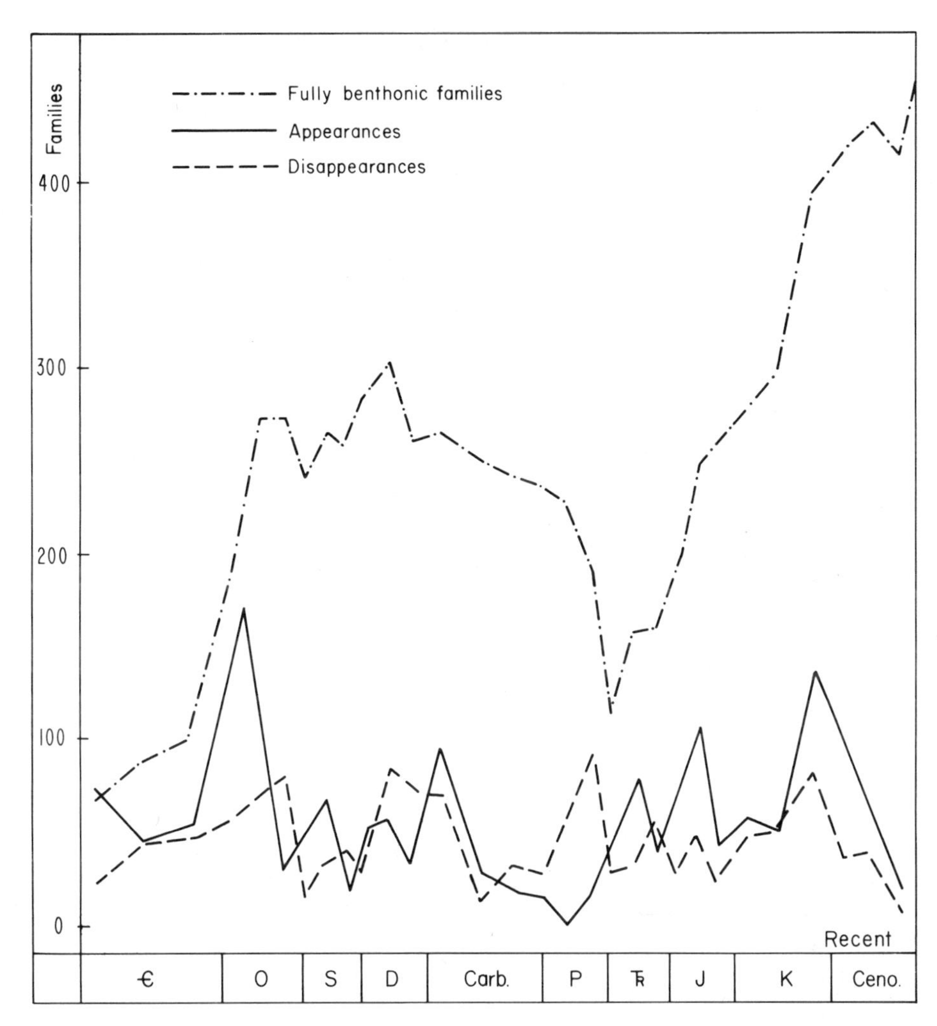

Figure 10–3: The Phanerozoic record of diversification, extinction, and standing diversity of the families of well-skeletonized benthic animals of the shelf seas belonging to the following phyla: Protozoa, Porifera, Archaeocyatha, Coelenterata, Brachiopoda, Ectoprocta, Mollusca, Arthropoda, and Echinodermata (from data presented in Valentine, 1969b).

with low-diversity states should contain a high proportion of trophic generalists, of detritus-feeders, of ubiquitous species, and in general of opportunistic species. And secondly, when the biosphere experiences a wave of extinction that is owing to decreased spatial heterogeneity of the environment, as when planetary temperature gradients are lowered or when barriers to dispersal are greatly decreased, diversity should again fall and remain at some lower level until further change, while provinciality should be concurrently reduced.

Diversity-independent factors, on the other hand, may cause waves of extinction whenever they change significantly on a planetary level. However, a wave of diversification would be expected to follow in a short time, geologically speaking, with the obvious exception of changes that cause some factor to assume a value that is generally lethal to metazoans. In this event natural selection would presumably have to operate for a relatively long time to create a whole new biota based on adaptation to the new situation. When density-independent factors operate there are no systematic changes required in the structure of the communities or in provinciality, and the accompanying waves of extinction will merely remove the taxa that happen to be poorly adapted to the new conditions.

In nature, most global changes in the environment will involve both diversity-dependent and diversity-independent factors. For example, changes in ocean temperature regimes will affect the strength of surface currents and of upwelling, changing the nutrient cycles and thereby affecting the trophic resource regime. Nevertheless, there are several events represented in *figure 10–3* that can be interpreted as being due chiefly to one or the other of these two modes of diversity regulation. The curves show a great wave of diversification in Cambro-Ordovician time accompanied by a wave of extinction which is so overshadowed that total diversity rises to a high level. Although other waves of extinction occur during the Paleozoic, they are also accompanied or are shortly followed by diversifications, so that total diversity is not much affected. It seems quite likely that these waves of extinction were caused chiefly by diversity-independent factors. However, the Permo-Triassic wave of extinction was neither accompanied nor closely followed by diversification, and therefore caused diversity levels to drop to a very low point. Triassic species tend to be cosmopolitan, and many are rather ubiquitous, and the more specialized lineages of the late Paleozoic largely disappeared. These observations suggest that diversity-dependent regulation was involved, and that both trophic and spatial regulators were operating.

The great Mesozoic and Cenozoic diversification that follows the Permo-Triassic diversity minimum is certainly not due to any optimization of conditions with respect to diversity-independent factors, but can be directly related to an increased potential for diversity that has accompanied an amelioration of diversity-dependent factors. Spatial environmental heterogeneity has increased greatly, permitting the rise of numerous provinces, partly owing to an increased latitudinal thermal gradient and partly to the separation of continents by sea-floor spreading and the appearance of dispersal barriers between them. Each province contains many endemic taxa, so that total diversity in the oceans has risen enormously during this provincialization, possibly by an order of magnitude (Valentine, 1969b).

Furthermore, the highly diverse ecosystems of the present tropics contain many more skeletonized species and many more specialists than any of the communities known from the early Mesozoic. It is therefore likely that a significant stabilization of trophic resources has occurred in low latitudes. Waves of extinction that are recorded during this general trend towards higher diversity, such as during the late Triassic and near the end of the Cretaceous, must be due to density-independent factors for the most part. Probably the most important density-independent factors are climatic.

Thus those major Phanerozoic events in the history of the biosphere that can be regarded as resulting from ecological diversity controls can be interpreted from our model of diversity regulation in a reasonable manner, although the model is still unproven. The model seems to be promising enough to follow up with subsequent studies and tests. Among the more important studies will be those that describe and interpret the structure and function of the invertebrate shelf ecosystems as they have evolved. As we learn more and more about the details of the evolution of ecosystems during the critical intervals of time, our ability to correctly interpret the causes of the major events in the history of the biosphere will clearly be much enhanced.

Summary and Implications of Ecosystem Modeling

In the opening sections we discussed briefly the reasons why conceptual models must be employed in working with ancient ecosystems and also the sources of ideas for such models. Then we erected a series of models purporting to explain how ecosystem structure and function might be regulated. These models, somewhat integrated into a hypothesis of diversity control, were then extended to account for anomalies, and expanded to account for the evolution of ecosystems, and applied to the major ecological events indicated by the fossil record. It is asserted that the models have explanatory power and may indicate the processes that actually operated in the past. It would be surprising if they were found to be correct in every respect; hardly anything can be made to work properly when it is first assembled.

The sources of ideas for the models can usually be identified. An especially important source for the diversity-regulation model is based on analogy with the controls on the number of individuals in a population. Thus density-dependence and -independence in populations are analogized with diversity-dependence and -independence in ecosystems, and the control of numbers of individuals in populations is compared with the control of numbers of populations in ecosystems. Such analogies may or may not be valid. This one appears to hold, not because it is a logical analogy, but because the trophic strategies of populations happen to affect their diversity. Ideas for defining the components of ecosystem evolution were also drawn from populations, in this case from the changes in the frequency (proportion, quality, and diversity) of genes within populations, which were analogized to changes in the frequency of the population niches within ecosystems. It would seem that in dealing with the evolution of ecological units, a likely source of helpful analogies is in other evolving systems, especially systems for which evolutionary processes are well known; in the present case it is the population level that is best known. Perhaps it is true generally of hierarchical models that analogy between hierarchical levels provides an especially important source of ideas. However, it is not a principle that processes on one level are similar to those on another, so that care must be exercised to avoid mistaken analogies.

During the discussion of the applications of the models to the fossil record it was tacitly assumed that the taxonomic composition of the ecosystems was of no special significance, and that the processes and trends predicted by the model were independent of the species that happened to inhabit the ecosystems. Different waves of extinction or of diversification were sometimes assigned similar causes, even though the fauna had changed very importantly. Is it really possible that in some situations we do not need to know too much about the species themselves, in order to make models of ecosystem evolution? This appears to be the case, and just why it is so deserves special notice, for many paleontologists are impressed with the uniqueness of taxa, the complexity of the biosphere, and the difficulty of dealing with the fossil record, and are loath to generalize and therefore to model.

When dealing with interactions that occur in a series of levels, as in a hierarchy, it is possible to "explain" the workings of a system on any given level in terms of the interactions between its primary subsystems alone; it is not necessary to understand the interior workings of those subsystems. All that is necessary is that an investigator make careful assumptions about the external properties of the primary subsystems; then they can be treated as "black boxes." The nature of the secondary subsystems that operate within the primary subsystem is of major interest if the workings of the primary subsystems are to be interpreted, but is not very critical if the workings of only the main system are under study. This property of near-independence of processes that operate within levels of a hierarchy is termed "near-decomposability" by Simon (1962), and has been applied to ecological units (Valentine, 1968). In an ecological hierarchy we are dealing with a sequence that proceeds from individuals to populations to communities to provinces to the biosphere. If we wish to investigate the functions of communities, that is to say, their ecosystems, we may do so in terms of the numbers and interactions of the populations that compose them, and need not be too concerned with the interactions of individuals or of characters, so long as we make correct assumptions about population interactions. These interactions may arise from processes among the characters, but that does not concern us. If an ecosystem is evolving towards trophic stability and if diversity is rising, then we do not care whether a given population is a crustacean or a gastropod, or whether its specialization involves modification of radula or mandible; we are only concerned that, say, trophic specialization occurs.

Of course if our interest were in the evolution of lineages rather than that of ecosystems, then the internal properties of the taxa would be all-important. The waves of extinction and diversification have certainly affected many taxonomic groups differentially, sweeping away whole classes and orders and raising others to a new dominance. Clearly, some taxa have been preadapted to some of the changes or to recovering well from them, while others have been all too vulnerable. Thus if rising trophic resource instabilities created a low-diversity state with few trophic specialists, the taxa that happened to contain highly specialized predators would be hard hit, while those made up chiefly of ubiquitous detritus-feeders would be favorably preadapted to the new regime, insofar as trophic resources were concerned. A return to a high-diversity state might lead to the diversification of species of a taxon that had formerly been rare, in order to replace the formerly abundant specialists. Thus, while the evolution of ecosystems is inextricably bound up with the evolution of taxa, it can nevertheless be studied in its own right. We do not need to wait for all the taxa in the fossil record to be described, or for all their distributional patterns to be recorded, in order to proceed with interpreting the history of ecosystems.

REFERENCES

Working scholars cannot live by reprints alone, but to use a library these days given the usual "information" in a journal citation (sometimes only euphemistically called a reference) is no mean feat. And to expect someone to obtain such "references" on inter-library loan is to expect the impossible, or at least the improbable. In contrast, we believe that the burden is on the author and the editor (and the publisher, which should not have a treasurer who deems a full reference "too expensive"), not on the unsuspecting reader. We have therefore taken this opportunity to give the titles of articles, and the full name of each journal. Mistakes in the citations in the published literature are common, and difficult to catch; we hope that those that may remain here will be more obvious than is usual, and hence less of a burden on the interested reader.

Alfvén, H. 1966. *Worlds-Antiworlds: Antimatter in Cosmology.* W. H. Freeman and Co. San Francisco. 103 pp.

Anonymous. 1967. *Did Man Get Here by Evolution or by Creation?* Watchtower Bible and Tract Society of Pennsylvania. New York. 191 pp.

Arnaud, P. M. 1970. Frequency and ecological significance of necrophagy among the benthic species of antarctic coastal waters. In *Antarctic Ecology.* M. W. Howdgate (Ed.). *1*:259–266. Academic Press. New York.

Arrhenius, O. 1921. Species and area. *Journal of Ecology. 9*:95–99.

Arrhenius, O. 1923. Statistical investigations in the constitution of plant associations. *Ecology. 4*:68–73.

Ax, P. 1966. Die Bedeutung der interstitiellen Sandfauna für allgemeine Probleme der Systematik, Ökologie und Biologie. *Veröffentlichungen des Instituts für Meeresforschung in Bremerhaven, Sonderband. 2*:15–66.

Babin, C. 1971. *Éléments de Paléontologie*. Armand Colin. Paris. 408 pp.

Bakus, G. J. 1967. The feeding habits of fishes and primary production at Eniwetok, Marshall Islands. *Micronesica. 3*:135–149.

Bassler, R. S. 1933. Development of Invertebrate Paleontology in America. *Geological Society of America Bulletin. 44*:265–286.

Beals, E. W. 1969. Vegetational change along altitudinal gradients. *Science. 165*:981–985.

de Beer, G. 1940. *Embryos and Ancestors*. Oxford University Press. Oxford. 108 pp. [Third Edition, 1958. 197 pp.]

de Beer, G. 1956. The evolution of Ratites. *Bulletin of the British Museum* (*Natural History*), (*Zoology*). *4*:59–70.

de Beer, G. 1970. The Evolution of Charles Darwin, review of M. T. Ghiselin, *The Triumph of the Darwinian Method. New York Review of Books. 15*:31–35.

Beerbower, J. R. 1960. *Search for the Past: an Introduction to Paleontology*. Prentice-Hall. Englewood Cliffs, New Jersey. 562 pp. [Second Edition, 1968.]

Belyaeva, N. V. 1963. Distribution of planktonic foraminifera on the floor of the Indian Ocean. *Voprosi Mikropaleontologii. 7*:209–222.

Belyaeva, N. V. 1964. Distribution of planktonic foraminifera in the water and on the floor of the Indian Ocean. *Akademiia Nauk S.S.S.R., Institut Okeanologii Trudy. 68*:12–83.

Berger, W. H. 1969. Planktonic Foraminifera: basic morphology and ecologic implications. *Journal of Paleontology. 43*:1369–1383.

Berggren, W. A. 1969. Rates of evolution in some Cenozoic planktonic foraminifera. *Micropaleontology. 15*:351–365.

Bernard, F. 1895. *Éléments de Paléontologie*. J.-B. Baillière and Son. Paris. 1168 pp. (Partly reprinted in pamphlet form, 91 pp., and translated into English; also said to be in the *Annual Report of the New York State Geological Survey. 14*:127–217.)

Berry, E. W. 1929. *Paleontology*. McGraw-Hill. New York. 392 pp.

Bevelander, G., and Benzer, P. 1948. Calcification in marine molluscs. *Biological Bulletin. 94*:176–183.

Beverton, R. J. H., and Holt, S. J. 1957. On the dynamics of exploited fish populations. *Fishery Investigations, London. 19*:1–533.

Blackwelder, R. E. 1964. Phyletic and phenetic *versus* omnispective classification. In *Phenetic and Phylogenetic Classification*. V. H. Heywood and J. McNeill (Eds.). The Systematics Association Publication No. 6. Systematics Association. London. pp. 17–28.

Blow, W. H. 1969. Late Middle Eocene to Recent planktonic foraminiferal biostratigraphy. In *Proceedings of the First International Conference of Planktonic Microfossils. Geneva. 1967*. E. J. Brill. Leiden. *1*:199–422.

Bock, W. J., 1963. The cranial evidence for ratite affinities. In *Proceedings, Thirteenth International Ornithological Congress, 1962. 1*:39–54.

Bock, W. J. 1969. Comparative morphology in systematics. In *Systematic Biology: Proceedings of an International Conference Conducted at the University of Michigan.* National Academy of Sciences. Publication No. 1692. Washington, D.C., pp. 411–458.

Bock, W. J., and von Wahlert, G. 1965. Adaptation and the form-function complex. *Evolution. 19*:269–299.

Boltovskoy, E. 1969. Living planktonic foraminifera at the 90°E meridian from the equator to the Antarctic. *Micropaleontology. 15*:237–255.

Boucot, A. J. 1953. Life and death assemblages among fossils. *American Journal of Science. 251*:25–40.

Boyden, A. 1935. Genetics and homology. *Quarterly Review of Biology. 10*:448–451.

Boyden, A. 1943. Homology and analogy. A century after the definitions of "homologue" and "analogue" of Richard Owen. *Quarterly Review of Biology. 18*:228–241.

Boyden, A. 1947. Homology and analogy. A critical review of the meanings and implications of these concepts in biology. *American Midland Naturalist. 37*:648–669.

Brace, C. L. 1967. *The Stages of Human Evolution.* Prentice-Hall. Englewood Cliffs, New Jersey. 116 pp.

Bretsky, P. W., and Lorenz, D. M. 1970. Adaptive response to environmental stability: a unifying concept in paleoecology. In *Proceedings of the North American Paleontological Convention, Chicago, 1969. E*:522–550. Allen Press. Lawrence, Kansas.

Broadhurst, F. M. 1964. Some aspects of the palaeoecology of non-marine faunas and rates of sedimentation in the Lancashire Coal Measures. *American Journal of Science. 262*:858–869.

Brouwer, A. 1968. *General Paleontology.* University of Chicago Press. Chicago. 216 pp. [Translated from the Dutch by R. H. Kaye.]

Brown, W. L., Jr. 1965. Numerical taxonomy, convergence, and evolutionary reduction. *Systematic Zoology. 14*:101–109.

Brown, W. L., Jr., and Wilson, E. O., 1956. Character displacement. *Systematic Zoology. 5*:49–64.

Burton, C. J. 1969. Variation studies of some phacopid trilobites of Eurasia and North West Africa. Ph.D. Thesis, University of Exeter, England. 222 pp.

Cadée, G. C. 1968. *Molluscan Biocoenoses and Thanatocoenoses in the Ria de Arosa, Galiocia, Spain.* E. J. Brill. Leiden. 121 pp.

Cairns, J., Dahlberg, M. L., Dickson, K. L., Smith, N., and Waller, W. T. 1969. The relationship of fresh-water protozoan communities to the

MacArthur-Wilson equilibrium model. *American Naturalist. 103*:439–454.

Callomon, J. H. 1963. Sexual dimorphism in Jurassic ammonites. *Transactions of the Leicester Literary and Philosophical Society. 57*:21–56.

Camin, J. H., and Sokal, R. R. 1965. A method for deducing branching sequences in phylogeny. *Evolution. 19*:311–326.

Cantlon, J. E. 1968. The continuum concept of vegetation: responses. *Botanical Review. 34*:253–258.

Carlquist, S. 1965. *Island Life: A Natural History of the Islands of the World.* Natural History Press. Garden City, New York. 451 pp.

Carruthers, R. G. 1910. On the evolution of *Zaphrentis delanouei* in Lower Carboniferous times. *Quarterly Journal of the Geological Society of London. 66*:523–538.

Chamberlain, J. A., Jr. 1969. Technique for scale modelling of cephalopod shells. *Palaeontology. 12*:48–55.

Chamberlain, J. A., Jr. 1971. Fluid mechanics of the ectocochliate cephalopod shell: an experimental study. Ph.D. thesis, Department of Geological Sciences, University of Rochester.

Chave, K. E. 1954. Aspects of the biogeochemistry of magnesium I. Calcareous marine organisms. *Journal of Geology. 62*:266–283.

Chave, K. E. 1964. Skeletal durability and preservation. In *Approaches to Paleoecology.* J. Imbrie and N. D. Newall (Eds.). John Wiley. New York. pp. 377–387.

Clark, R. B. 1964. *Dynamics in Metazoan Evolution: the Origin of the Coelom and Segments.* Oxford University Press. Oxford. 313 pp.

Clarke, F. W., and Wheeler, W. C. 1922. The inorganic constituents of marine invertebrates. *U.S. Geological Survey, Professional Paper 124.* 62 pp.

Clarkson, E. N. K. 1966. Schizochroal eyes and vision of some Silurian acastid trilobites. *Palaeontology. 9*:1–29.

Clements, F. E., and Shelford, V. E. 1939. *Bio-Ecology.* John Wiley. New York. 425 pp.

Clifton, H. E., and Boggs, S., Jr. 1970. Concave-up pelecypod (*Psephidia*) shells in shallow marine sand, Elk River Beds, Southwestern Oregon. *Journal of Sedimentary Petrology. 40*:888–897.

Cloud, P. E., Jr. 1968. Pre-metazoan evolution and the origins of the Metazoa. In *Evolution and Ecology.* E. T. Drake (Ed.). Yale University Press. New Haven, Conn. pp. 1–72.

Coe, W. R. 1957. Fluctuations in littoral populations. In *Treatise on Marine Ecology and Paleoecology.* J. W. Hedgpeth (Ed.). *Geological Society of America Memoir 67. 1*:935–939.

Cole, L. 1954. The population consequences of life history phenomena. *Quarterly Review of Biology. 29*:103–137.

Colless, D. H. 1969. Phylogenetic inference: a reply to Dr. Ghiselin. *Systematic Zoology. 18*:462–466.

Connell, J. H., Mertz, D. B., and Murdoch, W. W. 1970. *Readings in Ecology and Ecological Genetics.* Harper & Row. New York. 397 pp.

Connell, J. H., and Orias, E. 1964. The ecological regulation of species diversity. *American Naturalist. 98*:399–414.

Cowen, R. 1970. Analogies between the recent bivalve *Tridacna* and the fossil brachiopods *Lyttoniacea* and *Richthofeniacea. Palaeogeography, Palaeoclimatology, Palaeoecology. 8*:329–344.

Craig, G. Y. 1967. Size-frequency distributions of living and dead populations of pelecypods from Bimini, Bahamas, B.W.I. *Journal of Geology. 75*:34–45.

Craig, G. Y., and Hallam, A. 1963. Size-frequency and growth-ring analyses of *Mytilus edulis* and *Cardium edule*, and their palaeoecological significance. *Palaeontology. 6*:731–750.

Craig, G. Y., and Oertel, G. 1966. Deterministic models of living and fossil populations of animals. *Quarterly Journal of the Geological Society of London. 122*:315–355.

Culver, D. C. 1970. Analysis of simple cave communities. I. Caves as islands. *Evolution. 24*:463–474.

Darlington, P. J. 1957. *Zoogeography: The Geographical Distribution of Animals.* John Wiley. New York. 675 pp.

Darlington, P. J. 1965. *Biogeography of the Southern End of the World.* Harvard University Press. Cambridge, Mass. 236 pp.

Darlington, P. J. 1970. Carabidae on tropical islands, especially in the West Indies. *Biotropica. 2*:7–15.

Darwin, C. 1851–1854. *A Monograph on the Sub-class Cirripedia, with Figures of All the Species.* Volume 1, 400 pp.; volume 2, 684 pp. Ray Society. London.

Darwin, C. 1859. *On the Origin of Species by Means of Natural Selection, or the Preservation of Favoured Races in the Struggle for Life.* John Murray. London. 490 pp.

Darwin, C. 1877. *The Various Contrivances by Which Orchids are Fertilised by Insects.* Second edition. D. Appleton and Company. New York. 300 pp.

Davies, A. M. 1925. *An Introduction to Paleontology.* T. Murby & Co. London. 414 pp. [Second edition, 1947.]

Davis, D. D. 1949. Comparative anatomy and the evolution of vertebrates. In *Genetics, Paleontology and Evolution.* G. L. Jepsen, E. Mayr, and G. G. Simpson (Eds.). Princeton University Press. Princeton, New Jersey. pp. 61–89.

Deevey, E. S., Jr. 1947. Life tables for natural populations of animals, *Quarterly Review of Biology. 22*:283–314.

Degens, E. T., Johannesson, B. W., and Meyer, R. W. 1967. Mineralization processes in Molluscs and their paleontological significance. *Naturwissenschaften. 24*:638–640.

Diamond, J. M. 1969. Avifaunal equilibria and species turnover rates on the Channel Islands of California. *Proceedings of the National Academy of Sciences* (U.S.). *64*:57–63.

Dobzhansky, T. 1937. *Genetics and the Origin of Species.* Columbia University Press. New York. 364 pp.

Dodd, J. R. 1965. Environmental control of strontium and magnesium in *Mytilus. Geochimica et Cosmochimica Acta. 29*:385–398.

Dodd, J. R. 1967. Magnesium and strontium in calcareous skeletons: a review. *Journal of Paleontology. 41*:1313–1329.

Dogiel, V. A. 1966. *General Parasitology.* Academic Press. New York. 516 pp.

Dony, J. G. 1963. The expectation of plant records from prescribed areas. *Watsonia. 5*:377–385.

Douglas, A. 1970. Finite elements for geological modelling. *Nature. 226*:630–631.

Driscoll, E. G. 1967. Experimental field study of shell abrasion. *Journal of Sedimentary Petrology. 37*:1117–1123.

Driscoll, E. G. 1970. Selective bivalve shell destruction in marine environments, a field study. *Journal of Sedimentary Petrology. 40*:898–905.

Dunbar, M. J. 1968. *Ecological Development in Polar Regions.* Prentice-Hall. Englewood Cliffs, New Jersey. 119 pp.

Easton, W. H. 1960. *Invertebrate Paleontology.* Harper & Row. New York. 701 pp.

Eaton, T. H., Jr. 1970. *Evolution.* W. W. Norton. New York. 270 pp.

Ehrlich, P. R., and Raven, P. H. 1969. Differentiation of populations. *Science. 165*:1228–1232.

Eldredge, N. 1971. The allopatric model and phylogeny in Paleozoic invertebrates. *Evolution. 25*:156–167.

Eldredge, N. 1972. Systematics and evolution of *Phacops rana* (Green, 1832) and *Phacops iowensis* Delo, 1935 (Trilobita) in the middle Devonian of North American. *Bulletin of The American Museum of Natural History. 47*:45–114.

Elias, M. K. 1937. Depth distribution of the Big Blue (late Paleozoic) sediments in Kansas. *Geological Society of America Bulletin. 48*:403–432.

Epstein, S. 1959. The variations of the C^{18}/O^{16} ratio in nature and some geologic implications. In *Researches in Geochemistry.* P. H. Abelson (Ed.). John Wiley. New York. pp. 217–240.

Epstein, S., Buchsbaum, R., Lowenstam, H. A., and Urey, H. C. 1953. Revised carbonate-water isotopic temperature scale. *Geological Society of America Bulletin. 64*:1315–1326.

Epstein, S., and Mayeda, T. 1953. Variation of O^{18} content of waters from natural sources. *Geochimica et Cosmochimica Acta. 4*:213–224.

Fagerstrom, J. A. 1964. Fossil communities in paleoecology: their recognition and significance. *Geological Society of America Bulletin. 75*:1197–1216.

Feyerabend, P. K. 1970. Classical empiricism. In *The Methodological Heritage of Newton*. R. E. Butts and J. W. Davis (Eds.). University of Toronto Press. Toronto. pp. 150–170.

Fischer, A. G. 1960. Latitudinal variation in organic diversity. *Evolution. 14*:64–81.

Fischer, A. G., and Garrison, R. E. 1967. Carbonate lithification on the sea floor. *Journal of Geology. 75*:488–496.

Fitch, W. M. 1970. Distinguishing homologous from analogous proteins. *Systematic Zoology. 19*:99–113.

Fitch, W. N., and Margoliash, E. 1967. Construction of phylogenetic trees. *Science. 155*:279–284.

Fryer, G., and Iles, T. D. 1969. Alternative routes to evolutionary success as exhibited by African cichlid fishes of the genus *Tilapia* and the species flocks of the Great Lakes. *Evolution. 23*:359–369.

Fyfe, W. S., and Bischoff, J. L. 1965. The calcite-aragonite problem. In *Dolomitization and Limestone Diagenesis*. L. C. Pray and R. C. Murray (Eds.). Society of Economic Paleontologists and Mineralogists, Special Publication 13. pp. 3–13.

Ghiselin, M. T. 1966a. An application of the theory of definitions to systematic principles. *Systematic Zoology. 15*:127–130.

Ghiselin, M. T. 1966b. On psychologism in the logic of taxonomic controversies. *Systematic Zoology. 15*:207–215.

Ghiselin, M. T. 1966c. Reproductive function and the phylogeny of opisthobranch gastropods. *Malacologia. 3*:327–378.

Ghiselin, M. T. 1969a. *The Triumph of the Darwinian Method*. University of California Press. Berkeley and Los Angeles. 287 pp.

Ghiselin, M. T. 1969b. The distinction between similarity and homology. *Systematic Zoology. 18*:148–149.

Ghiselin, M. T. 1969c. The evolution of hermaphroditism among animals. *Quarterly Review of Biology. 44*:189–208.

Ghiselin, M. T. 1971. The individual in the Darwinian Revolution. *New Literary History. 3*:113–134.

Gleason, H. A. 1922. On the relation between species and area. *Ecology. 3*:158–162.

Gleason, H. A. 1925. Species and area. *Ecology. 6*:66–74.

Gleason, H. A. 1939. The individualistic concept of the plant association. *American Midland Naturalist. 21*:92–110.

Glimcher, M. J. 1960. Specificity of the molecular structure of organic matrices in mineralisation. In *Calcification in Biological Systems.* R. F. Sognnales (Ed.). American Association for the Advancement of Science, Publication No. 64. pp. 421–487.

Gooch, J. L., and Schopf, T. J. M. 1970. Population genetics of marine species of the Phylum Ectoprocta. *Biological Bulletin. 138*:138–156.

Gould, S. J. 1965. Is uniformitarianism necessary? *American Journal of Science. 263*:223–228.

Gould, S. J. 1969. An evolutionary microcosm: Pleistocene and Recent history of the land snail *P.* (*Poecilozonites*) in Bermuda. *Bulletin of the Museum of Comparative Zoology. 138*:407–531.

Gould, S. J. 1970a. Evolutionary paleontology and the science of form. *Earth-Science Reviews. 6*:77–119.

Gould, S. J. 1970b. Coincidence of climatic and faunal fluctuations in Pleistocene Bermuda. *Science. 168*:572–573.

Gould, S. J. 1970c. Dollo on Dollo's law: irreversibility and the status of evolutionary laws. *Journal of the History of Biology. 3*:189–212.

Gould, S. J. 1971a. D'Arcy Thompson and the science of form. *New Literary History. 2*:229–258.

Gould, S. J. 1971b. Geometric similarity in allometric growth: a contribution to the problem of scaling in the evolution of size. *American Naturalist. 105*:113–136.

Gould, S. J. 1971c. Tübingen meeting on form. *Journal of Paleontology. 45*:1042–1043.

Gould, S. J. *In press.* Allometric fallacies and the evolution of *Gryphaea*: a new interpretation based on White's criterion of geometric similarity. *Evolutionary Biology.*

Grant, P. R. 1966. Ecological compatibility of bird species on islands. *American Naturalist. 100*:451–462.

Grant, P. R. 1968. Bill size, body size, and the ecological adaptations of bird species to competitive situations on islands. *Systematic Zoology. 17*:319–333.

Greene, M. 1958. Two evolutionary theories. Part I. *The British Journal for the Philosophy of Science. 9*:110–127. Part II. *The British Journal for the Philosophy of Science. 9*:185–193.

Greenslade, P. J. M. 1968a. The distribution of some insects of the Solomon Islands. *Proceedings of the Linnaean Society of London. 179*:189–196.

Greenslade, P. J. M. 1968b. Island patterns in the Solomon Islands bird fauna. *Evolution. 22*:751–761.

Greenslade, P. J. M. 1969. Insect distribution patterns in the Solomon Islands. *Philosophical Transactions of the Royal Society of London, Series B. 255*:271–285.

Gunter, G. 1957. Temperature. In *Treatise on Marine Ecology and Paleoecology*. J. W. Hedgpeth (Ed.). *Geological Society of America Memoir 67*. *1*:359–382.

Haas, O., and Simpson, G. G. 1946. Analysis of some phylogenetic terms, with attempts at redefinition. *Proceedings of the American Philosophical Society*. *90*:319–349.

Hagmeier, E. M., and Stults, C. D. 1964. A numerical analysis of the distributional patterns of North American mammals. *Systematic Zoology* *13*:125–155.

Haldane, J. B. S. 1932. *The Causes of Evolution*. Longmans, Green & Co., Ltd. London, 235 pp. [Cornell University Press. Cornell Paperbacks. 1966.]

Hallam, A. 1959. On the supposed evolution of *Gryphaea* in the Lias. *Geological Magazine*. *96*:99–108.

Hallam, A. 1961. Brachiopod life assemblages from the Marlstone Rockbed of Leicestershire. *Palaeontology*. *4*:653–659.

Hallam, A. 1962. The evolution of *Gryphaea*. *Geological Magazine*. *99*:571–574.

Hallam, A. 1965. Environmental causes of stunting in living and fossil marine benthonic invertebrates. *Palaeontology*. *8*:132–155.

Hallam, A. 1967. The interpretation of size-frequency distributions in molluscan death assemblages. *Palaeontology*. *10*:25–42.

Hallam, A. 1968. Morphology, paleoecology, and evolution of the genus *Gryphaea* in the British Lias. *Philosophical Transactions*, *Royal Society of London*, *Series B*. *254*:91–128.

Hallam, A. 1971. A facies analysis of the Lias in Portugal. *Neues Jahrbuch für Mineralogie, Geologie, und Paläontologie, Abteilung*. *139*:226–265.

Hallam, A., and Price, N. B. 1968. Environmental and biochemical control of strontium in shells of *Cardium edule*. *Geochimica et Cosmochimica Acta*. *32*:319–328.

Hamilton, T. H., Barth, R. H., and Rubinoff, I. 1964. The environmental control of insular variation in bird species abundance. *Proceedings of the National Academy of Science* (U.S.) *52*:132–140.

Hamilton, T. H., and Rubinoff, I. 1967. On predicting insular variation in endemism and sympatry for the Darwin Finches in the Galápagos Archipelago. *American Naturalist*. *101*:161–171.

Hancock, D. A., and Simpson, A. C. 1961. Parameters of marine invertebrate populations. In *The Exploitation of Natural Animal Populations*. E. D. Le Cren and M. W. Holdgate (Eds.). John Wiley. New York. pp. 29–50.

Hanson, E. D. 1963. Homologies and the Ciliate origin of the Eumetazoa. In *The Lower Matazoa: Comparative Biology and Phylogeny*. E. C. Dougherty (Ed.). University of California Press. Berkeley and Los Angeles. pp. 7–22.

Hanson, N. R. 1969. *Perception and Discovery: An Introduction to Scientific Inquiry*. Freeman, Cooper and Co. San Francisco. 435 pp.

Hanson, N. R. 1970. Hypotheses fingo. In *The Methodological Heritage of Newton*. R. E. Butts and J. W. Davis (Eds.). University of Toronto Press. Toronto. pp. 14–33.

Harbaugh, J. W., and Bonham-Carter, G. 1970. *Computer Simulation in Geology*. John Wiley. New York. 575 pp.

Hare, P. E. 1963. Amino acids in the protein from aragonite and calcite in the shells of *Mytilus californianus*. *Science*. *139*:216–217.

Harland, W. B., *et al.* (Eds.). 1967. *The Fossil Record*. Geological Society of London. London. 828 pp.

Harrison, G. A., and Weiner, J. S. 1963. Some considerations in the formulation of theories of human phylogeny. In *Classification and Human Evolution*. S. L. Washburn (Ed.). Aldine. Chicago. pp. 75–84.

Helmcke, J. G., and Otto, F. 1962. Lebende und technische Konstruktionen. *Deutsche Bauzeitung, Jahrg. 67. 11*:855–861.

Hennig, W. 1950. *Grundzüge einer Theorie der phylogenetischen Systematik*. Deutscher Zentralverlag. Berlin. 370 pp. English translation, 1966. *Phylogenetic Systematics*. University of Illinois Press. Urbana, Illinois. 263 pp.

Hertel, H. 1966. *Structure-Form-Movement*. Reinhold Publishing Co. New York. 251 pp.

Hertwig, R. 1914. Die Abstammungslehre. In *Die Kultur der Gegenwart*. P. Hinneberg (Ed.), Part III, Mathematik, Näturwissenschaften, Medizin. Section 4, Organische Naturwissenschaften. Volume 4, Abstammungslehre, Systematik, Paläontologie, Biogeographie, under the direction of R. Hertwig and R. v. Wettstein. B. G. Teubner. Leipzig. pp. 1–91.

Hessler, R. R., and Sanders, H. L. 1967. Faunal diversity in the deep-sea. *Deep-Sea Research*. *14*:65–79.

Hildebrand, J. H. 1964. The use of models in physical science. *Proceedings of the American Philosophical Society*. *108*:411–417.

Holland, H. D., Kirsipu, T. V., Heubner, J. S., and Oxburgh, U. M. 1964. On some aspects of the chemical evolution of cave waters. *Journal of Geology*. *72*:36–67.

Holland, P. W. 1968. Some properties of a biogeography model. *Memorandum NS-107*. Department of Statistics, Harvard University.

Horowitz, A. S., and Potter, P. E. 1971. *Introductory Petrography of Fossils*. Springer-Verlag. New York. 302 pp.

Howell, F. C. 1967. Recent advances in human evolutionary studies. *Quarterly Review of Biology. 42*:471–513.

Howell, T. R. 1969. Avian distribution in Central America. *The Auk. 86*:293–326.

Hubbard, J. A. E. B. 1970. Sedimentological factors affecting the distribution and growth of Viséan caninioid corals in north-west Ireland. *Palaeontology. 13*:191–209.

Hubbard, M. D. 1971. The applicability of the MacArthur-Wilson species equilibrium model to artificial ponds. Master's thesis, Department of Biological Sciences, Florida State University.

Hull, D. L. 1964. Consistency and monophyly. *Systematic Zoology. 13*:1–11.

Hull, D. L. 1967. Certainty and circularity in evolutionary taxonomy. *Evolution. 21*:174–189.

Hutchins, L. W. 1947. The bases for temperature zonation in geographical distribution. *Ecological Monographs. 17*:325–335.

Imbrie, J. 1957. The species problem with fossil animals. In *The Species Problem*. E. Mayr (Ed.). American Association for the Advancement of Science, Publication No. 50. pp. 125–153.

Inger, R. F. 1958. Comments on the definition of genera. *Evolution. 12*:370–384.

Inglis, W. G. 1966. The observational basis of homology. *Systematic Zoology. 15*:219–228.

Janzen, D. H. 1968. Host plants as islands in evolutionary and contemporary time. *American Naturalist. 102*:592–595.

Jardine, N. 1969. The observational and theoretical components of homology: a study based on the morphology of the dermal skull-roofs of rhiphistian fishes. *Biological Journal of the Linnaean Society* [of London]. *1*:321–361.

Jeffries, R. P. S., and Minton, P. 1965. The mode of life of two Jurassic species of *Posidonia* (Bivalvia). *Palaeontology. 8*:156–185.

Jepsen, G. L., Mayr, E., and Simpson, G. G. (Eds.). 1949. *Genetics, Paleontology and Evolution*. Princeton University Press. Princeton, New Jersey. 474 pp.

Johnson, M. P., Mason, L. G., and Raven, P. H. 1968. Ecological parameters and plant species diversity. *American Naturalist. 102*:297–306.

Johnson, M. P., and Raven, P. H. 1970. Natural regulation of plant species diversity. *Evolutionary Biology. 4*:127–162.

Johnson, R. G. 1965. Pelecypod death assemblages in Tomales Bay, California. *Journal of Paleontology. 39*:80–85.

Johnson, R. G. 1970. Variations in diversity within benthic marine communities. *American Naturalist. 104*:285–300.

Johnson, R. G. 1971. Animal-sediment relations in shallow water benthic communities. *Marine Geology. 11*:93–104.

Juskevice, J. A. 1969. Interspecific correlation and association in benthic marine communities. Ph.D. thesis, Department of Geology, University of Chicago.

Kac, M. 1969. Some mathematical models in science. *Science. 166*:695–699.

Kaufmann, K. W. 1970. A model for predicting the influence of colony morphology on reproductive potential in the Phylum Ectoprocta. *Biological Bulletin. 139*:426.

Kaufmann, K. W. 1971. The effect of colony morphology on the life history strategy of bryozoans. *Abstracts with Programs.* Washington, D.C. 1971 Annual Meetings. The Geological Society of America. *3*:618.

Kellogg, V. L. 1913. Distribution and species-forming of ecto-parasites. *American Naturalist. 47*:129–158.

Kendeigh, S. C. 1961. *Animal Ecology.* Prentice-Hall. Englewood Cliffs, New Jersey. 468 pp.

Kinne, O. 1963. The effects of temperature and salinity on marine and brackish water animals. I. Temperature. *Oceanography and Marine Biology, Annual Review. 1*:301–340.

Kinsman, D. J. J., and Holland, H. D. 1969. The co-precipitation of cations with $CaCO_3$. IV. The co-precipitation of Sr^{2+} with aragonite between 16° and 96°C. *Geochimica et Cosmochimica Acta. 33*:1–17.

Kitano, Y. 1962. The behavior of various inorganic ions in the separation of calcium carbonate from a bicarbonate solution. *Bulletin of the Chemical Society of Japan. 35*:1973–1980.

Kitano, Y., and Hood, D. W. 1965. The influence of organic material on the polymorphic crystallization of calcium carbonate. *Geochimica et Cosmochimica Acta. 29*:29–41.

Kitts, D. B. 1963. The theory of geology. In *The Fabric of Geology.* C. C. Albritton, Jr. (Ed.). Freeman, Cooper & Company. San Francisco. pp. 49–68.

Klevezal', G. A., and Kleinenberg, S. E. 1969. Age determination of mammals from annual layers in teeth and bones. Israel Program for Scientific Translations. Jerusalem. 128 pp. [Opredeleniye Bozrasta Mlekopitayuschchikh po Sloistym Structuram Zubov i Kosti. Izdatel'stvo "Nauka"; Moscow. 1967.]

Klopfer, P. H., and MacArthur, R. H. 1960. Niche size and faunal diversity. *American Naturalist. 94*:293–300.

Kobayashi, S. 1964. Studies of shell formation. X. A study of the proteins of the extrapallial fluid in some molluscan species. *Biological Bulletin. 126*:414–422.

Kohn, A. J., and Helfrich, P. 1957. Primary organic productivity of a Hawaiian coral reef. *Limnology and Oceanography. 2*:241–251.

Koyré, A. 1968. *Newtonian Studies*. University of Chicago Press. Chicago. 288 pp.

Kuhn, T. S. 1962. *The Structure of Scientific Revolutions*. University of Chicago Press. Chicago. 172 pp. [Second edition 1970, 210 pp.]

Kullmann, J., and Scheuch, J. 1970. Wachstums-Änderungen in der Ontogenese paläozoischer Ammonoideen. *Lethaia*. *3*:397–412.

Kummel, B., and Lloyd, R. M. 1955. Experiments on relative streamlining of coiled cephalopod shells. *Journal of Paleontology*. *29*:159–170.

Kurtén, B. 1964. The population dynamic approach to paleoecology. In *Approaches to Paleoecology*. J. Imbrie and N. D. Newell (Eds.). John Wiley. New York. pp. 91–106.

Kurtén, B. 1965. Evolution in geological time. In *Ideas in Modern Biology*. J. A. Moore (Ed.). Natural History Press. Garden City, New York. pp. 329–354.

Lack, D. 1947. *Darwin's Finches*. Cambridge University Press. Cambridge. 208 pp.

Lam, H. J. 1936. Phylogenetic symbols, past and present. *Acta Biotheoretica*. *2*:153–194.

Latimer, W. M. 1952. *Oxidation Potentials*. Second Edition. Prentice-Hall. Englewood Cliffs, New Jersey. 392 pp.

Lehmann, U. 1966. Dimorphismus bei Ammoniten der Ahrensburger Lias-Geschiebe. *Paläontologische Zeitschrift*. *40*:26–55.

Leopold, L. B. 1969. Quantitative comparison of some aesthetic factors among rivers. *U.S. Geological Survey, Circular No. 620*. 16 pp.

Lerman, A. 1965a. Paleoecological problems of Mg and Sr in biogenic calcites in light of recent thermodynamic data. *Geochimica et Cosmochimica Acta*. *29*:977–1002.

Lerman, A. 1965b. Strontium and magnesium in water and in *Crassostrea* calcite. *Science*. *150*:745–751.

Lerner, I. M. 1954. *Genetic Homeostasis*. John Wiley. New York. 134 pp.

Lever, J., van den Bosch, M., Cook, H., van Dijk, T., Thiadens, A. J. H., and Thijssen, R. 1964. Quantitative beach research: III. An experiment with artificial valves of *Donax vittatus*. *Netherlands Journal of Sea Research*. *2*:458–492.

Levins, R. 1966. The strategy of model building in population biology. *American Scientist*. *54*:421–431.

Levins, R. 1968. *Evolution in Changing Environments*. Princeton University Press. Princeton, New Jersey. 120 pp.

Levins, R. 1969. Some demographic and genetic consequences of environmental heterogeneity for biological control. *Entomological Society of America Bulletin*. *15*:237–240.

Levinton, J. S. 1970. The paleoecological significance of opportunistic species. *Lethaia*. *3*:69–78.

Levinton, J. S., and Bambach, R. K. 1970. Some ecological aspects of bivalve mortality patterns. *American Journal of Science*. *268*:97–112.

Lewontin, R. C. 1957. The adaptation of populations to varying environments. *Cold Spring Harbor Symposia on Quantitative Biology*. *22*:395–408.

Lewontin, R. C. 1958. Studies on heterozygosity and homeostasis II: Loss of heterosis in a constant environment. *Evolution*. *12*:494–503.

Lewontin, R. C. 1963. Models, mathematics and metaphors. *Synthese*. *15*:222–244.

Lewontin, R. C. 1965. Selection for colonising ability. In *The Genetics of Colonising Species*. H. G. Baker and G. L. Stebbins (Eds.). Academic Press. New York. pp. 77–91.

Lewontin, R. C. 1969a. The bases of conflict in biological explanation. *Journal of the History of Biology*. *2*:35–45.

Lewontin, R. C. 1969b. The meaning of stability. *Brookhaven Symposia in Biology*. *22*:13–24.

Lie, U. 1968. A quantitative study of benthic infauna in Puget Sound. *Fiskeridirektoratets Skrifter, Havundersøkelser*. *14*:229–556.

Lindroth, A. 1935. Die Association der marinen Weichboden. *Zoologiska Bidrag från Uppsala*. *15*:331–336.

Lloyd, R. M. 1964. Variations in the oxygen and carbon isotope ratios of Florida Bay mollusks and their environmental significance. *Journal of Geology*. *72*:84–111.

Lowenstam, H. A. 1954a. Factors affecting the aragonite:calcite ratios in carbonate secreting marine organisms. *Journal of Geology*. *62*:284–322.

Lowenstam, H. A. 1954b. Environmental relations of modification compositions of certain carbonate secreting marine invertebrates. *Proceedings of the National Academy of Sciences* (U.S.). *40*:39–48.

Lowenstam, H. A. 1961. Mineralogy, O^{18}/O^{16} ratios, and strontium and magnesium contents of recent and fossil brachiopods and their bearing on the history of the oceans. *Journal of Geology*. *69*:241–260.

Lowenstam, H. A. 1962. Magnetite in denticle capping in Recent chitons (Polyplacophora). *Geological Society of America Bulletin*. *73*:435–438.

Lowenstam, H. A. 1963. Biologic problems relating to the composition and diagenesis of sediments. In *The Earth Sciences: Problems and Progress in Current Research*. T. W. Donnelly (Ed.). The University of Chicago Press. Chicago. pp. 137–195.

Lowenstam, H. A. 1964. Sr/Ca ratio of skeletal aragonites from the recent marine biota at Palau and from fossil gastropods. In *Isotopic and Cosmic Chemistry*. H. Craig, S. L. Miller, and G. J. Wasserburg (Eds.). North Holland Publishing Company. Amsterdam. pp. 114–132.

MacArthur, R. H. 1955. Fluctuations of animal populations and a measure of stability. *Ecology*. *36*:533–536.

MacArthur, R. H. 1957. On the relative abundance of bird species. *Proceedings of the National Academy of Sciences* (U.S.). *43*:293–295.

MacArthur, R. H. 1960. On the relative abundance of species. *American Naturalist*. *94*:25–36.

MacArthur, R. H. 1972. *Geographical Ecology*. Harper & Row. New York. 269 pp.

MacArthur, R. H., and Wilson, E. O. 1963. An equilibrium theory of insular zoogeography. *Evolution*. *17*:373–387.

MacArthur, R. H., and Wilson, E. O. 1967. *The Theory of Island Biogeography*. Princeton University Press. Princeton, New Jersey. 203 pp.

MacDonald, G. J. F. 1956. Experimental determination of calcite-aragonite equilibrium relations at elevated temperatures and pressures. *American Mineralogist*. *41*:744–756.

MacGillavry, H. J. 1968. Modes of evolution mainly among marine invertebrates. *Bijdragen tot de dierkunde*. *38*:69–74.

Mach, E. 1893. *The Science of Mechanics*. Open Court Publishing Co. La Salle, Illinois. 534 pp. [Translated from the second edition by T. J. McCormack.]

Maguire, B. 1963. The passive dispersal of small aquatic organisms and their colonization of isolated bodies of water. *Ecological Monographs*. *33*:161–185.

Malyshev, L. I. 1969. The dependence of the species abundance of a flora on environmental and historical factors. *Botanical Journal* (*Academy of Sciences, S.S.S.R.*). *54*:1137–1147.

Mandelbaum, M. 1964. *Philosophy, Science and Sense Perception: Historical and Critical Studies*. The Johns Hopkins Press. Baltimore. 262 pp.

Mangum, C. P. 1964. Studies on speciation in maldanid polychaetes of the North Atlantic coast. II. Distribution and competitive interaction of five sympatric species. *Limnology and Oceanography*. *9*:12–26.

Marcus, E. du B. R., and Marcus, E. 1960. On *Siphonaria hispida*. *Boletim do Faculdade de Filosofia, Ciências e Letras. Universidade de São Paulo,* (*Zoologia*). *23*:107–139.

Margalef, R. 1968. *Perspectives in Ecological Theory*. University of Chicago Press. Chicago. 111 pp.

Marsh, J. A., Jr. 1970. Primary productivity of reef-building calcareous red algae. *Ecology*. *51*:255–263.

Martens, C. S., and Harriss, R. C. 1970. Inhibition of apatite precipitation in the marine environment by magnesium ions. *Geochimica et Cosmochimica Acta*. *34*:621–625.

Maslin, T. P. 1952. Morphological criteria of phyletic relationships. *Systematic Zoology*. *1*:49–70.

Matthew, W. D. 1915. Climate and evolution. *Annals of the New York Academy of Sciences*. *24*:171–318.

Matthew, W. D., and Chubb, S. H. 1921. *Evolution of the Horse*. American Museum of Natural History, Guide Leaflet no. 36. 67 pp.

Mayr, E. 1942. *Systematics and the Origin of Species*. Columbia University Press. New York. 334 pp.

Mayr, E. 1959. Isolation as an evolutionary factor. *Proceedings of the American Philosophical Society*. *103*:221–230.

Mayr, E. 1963. *Animal Species and Evolution*. Harvard University Press. Cambridge, Mass. 797 pp.

Mayr, E. 1965a. Avifauna: Turnover on islands. *Science*. *150*:1587–1588.

Mayr, E. 1965b. Numerical phenetics and taxonomic theory. *Systematic Zoology*. *14*:73–97.

Mayr, E. 1969. *Principles of Systematic Zoology*. McGraw-Hill. New York. 428 pp.

Mayr, E. 1970. *Populations, Species and Evolution*. Harvard University Press. Cambridge, Mass. 453 pp.

McAlester, A. L. 1962. Some comments on the species problem. *Journal of Paleontology*. *36*:1377–1381.

McCrea, J. M. 1950. On the isotopic chemistry of carbonates and a paleotemperature scale. *Journal of Chemical Physics*. *18*:849–857.

McKerrow, W. S., Johnson, R. T., and Jakobson, M. E. 1969. Palaeoecological studies in the Great Oolite at Kirtlington, Oxfordshire. *Palaeontology*. *12*:56–83.

Medawar, P. B. 1967. *The Art of the Soluble*. Methuen & Co. London. 160 pp.

Medawar, P. B. 1969. *Induction and Intuition in Scientific Thought*. American Philosophical Society. Philadelphia. 62 pp.

Meise, W. 1963. Verhalten der Straussartigen Vögel und Monophylie der Ratite. In *Proceedings, Thirteenth International Ornithological Congress, 1962*. *1*:115–125.

Merkt, J. 1966. Über Austern und Serpeln als Epöken auf Ammonitengehäusen. *Neues Jahrbuch für Geologie und Paläontologie, Abhandlungen*. *125*:467–479.

Merriam, C. H. 1894. The geographic distribution of animals and plants in North America. U.S. Department of Agriculture, *Yearbook*, for 1894. pp. 203–214.

Merriam, C. H., Bailey, V., Nelson, E. W., and Preble, E. A. 1910. *Fourth Provisional Zone Map of North America*. Geological Survey, Biological Survey Bureau. U.S. Department of Agriculture. [Reprinted 1923.]

Merton, R. K. 1965. *On the Shoulders of Giants*. Harcourt, Brace and World. New York. 290 pp.

Milliman, J. D., Gastner, M., and Müller, J. 1971. Utilization of Magnesium in Coralline Algae. *Geological Society of America Bulletin*. *82*:573–580.

Mills, E. L. 1969. The community concept in marine zoology, with comments on continua and instability in some marine communities: a review. *Journal of the Fisheries Research Board of Canada. 26*:1415–1428.

Moore, D. G., and Scruton, P. C. 1957. Minor internal structures of some recent unconsolidated sediments. *American Association of Petroleum Geologists Bulletin. 41*:2723–2751.

Moore, H. B. 1958. *Marine Ecology*. John Wiley. New York. 493 pp.

Moore, P. F. 1967. The use of geological models in prospecting for stratigraphic traps. In *Proceedings, Seventh World Petroleum Congress, Mexico, 1967. 2*:481–485. Elsevier. London.

Moore, R. C., Lalicker, C. G., and Fischer, A. G. 1952. *Invertebrate Fossils*. McGraw-Hill. New York. 766 pp.

Moore, R. C., *et al.* 1968. Developments, trends, and outlooks in Paleontology. *Journal of Paleontology. 42*:1327–1377.

Murray, J. W. 1954. The deposition of calcite and aragonite in caves. *Journal of Geology. 62*:481–492.

Muscatine, L., and Cernichiari, E. 1969. Assimilation of photosynthetic products of Zooxanthellae by a reef coral. *Biological Bulletin. 137*:506–523.

Nabokov, V. 1969. *Ada or Ardor: A Family Chronicle*. McGraw-Hill. New York. 445 pp.

Naef, A. 1919. *Idealistische Morphologie und Phylogenetik*. Fischer. Jena. 77 pp.

Naef, A. 1927. Die Definition des Homologiebegriffes. *Biologisches Zentrelblatt. 47*:187–190.

Naef, A. 1931. Die Gestalt als Begriff und Idee. *Handbuch der vergleichenden Anatomie der Wirbeltiere. 1*:77–118.

Neef, G. 1970. Notes on the subgenus *Pelicaria*. *New Zealand Journal of Geology and Geophysics. 13*:436–476.

Nelson, D. J. 1964. Deposition of strontium in relation to morphology of clam (Unionidae) shells. *Verhandlungen der Internationalen Vereinigung der Theoretische und Angewante Limnologie. 15*:893–902.

Neurath, H., Walsh, K. A., and Winter, W. P. 1967. Evolution of structure and function of proteases. *Science. 158*:1638–1644.

Neville, A. C. 1967. Daily growth layers in animals and plants. *Biological Reviews. 42*:421–441.

Newell, N. D. 1954. Toward a more ample invertebrate paleontology. In Status of Invertebrate Paleontology, 1953. B. Kummel (Ed.). *Bulletin of the Museum of Comparative Zoology. 112*:93–97.

Newell, N. D. 1956. Catastrophism and the fossil record. *Evolution. 10*:97–101.

Newell, N. D. 1967. Revolutions in the history of life. *Geological Society of America, Special Papers. 89*:63–91.

Newton, I. 1726. *Philosophiae Naturalis Principia Mathematica.* Guil. & Joh. Innys, Pub. [Reprinted edition of 1871. J. Maclehose, Pub. Glasgow. 538 pp.]

Nichols, D. 1959. Changes in the chalk heart-urchin *Micraster* interpreted in relation to living forms. *Philosophical Transactions of the Royal Society of London. Series B. 242*:347–437.

Nikol'skiĭ, G. V. 1969. *Theory of Fish Population Dynamics.* R. Jones (Ed.). Translated by J. E. Bradley. S. H. Service Agency, Inc. New York. 323 pp.

Odum, H. T. 1951. Notes on the strontium content of sea water, celestite radiolaria, and strontianite snail shells. *Science. 114*:211–213.

Odum, H. T. 1957a. Strontium in natural waters. *Publication of the Institute of Marine Science. 4*:22–37.

Odum, H. T. 1957b. Biogeochemical deposition of strontium. *Publication of the Institute of Marine Science. 4*:38–114.

Odum, H. T. 1960. Ecological potential and analogue circuits for the ecosystem. *American Scientist. 48*:1–8.

Odum, H. T., and Odum, E. P. 1955. Trophic structure and productivity of a windward coral reef community on Eniwetok Atoll. *Ecological Monographs. 25*:291–320.

Olson, E. C. 1957. Size-frequency distributions in samples of extinct organisms. *Journal of Geology. 65*:309–333.

Olson, E. C., and Miller, R. L. 1958. *Morphological Integration.* University of Chicago Press. Chicago. 317 pp.

d'Orbigny, A. 1849–1852. *Cours Élémentaire de Paléontologie et de Géologie Stratigraphiques.* V. Masson. Paris. 3 volumes. 848 pp.

Osborn, H. F. 1905. The present problems of paleontology. *Popular Science Monthly.* pp. 226–242.

Owen, R. 1848. *On the Archetype and Homologies of the Vertebrate Skeleton.* Taylor & Taylor. London. 203 pp.

Pannella, G., and MacClintock, C. 1968. Biological and environmental rhythms reflected in molluscan shell growth. *Journal of Paleontology, Memoir. 2*:64–80.

Pantin, C. F. A. 1966. Homology, analogy and chemical identity in the Cnidaria. *Symposia of the Zoological Society of London. 16*:1–17.

Patrick, R. 1970. Benthic stream communities. *American Scientist. 58*:546–549.

Peters, J. A. 1968. A computer program for calculating degree of biogeographical resemblance between areas. *Systematic Zoology. 17*:64–69.

Pianka, E. R. 1966. Latitudinal gradients in species diversity: a review of concepts. *American Naturalist. 100*:33–46.

Pictet, F. J. 1853–1857. *Traité de Paléontologie ou Histoire Naturelle des Animaux Fossiles Considerés dan leurs Rapport Zoologiques et Géologiques.* Second edition. Paris. 4 volumes and atlas. [First edition, 1844–1846.]

Pilbeam, D. R. 1968. Human origins. *Advancement of Science. 24*:368–378.

Pilbeam, D. R., and Simons, E. L. 1965. Some problems of hominid classification. *American Scientist. 53*:237–259.

Pilkey, O. H., and Hower, J. 1960. The effect of environment on the concentration of skeletal magnesium and strontium in *Dendraster. Journal of Geology. 68*:203–216.

Platt, J. R. 1964. Strong inference. *Science. 146*:347–353.

Preston, F. W. 1962. The canonical distribution of commonness and rarity. Part I. *Ecology. 43*:185–215. Part II. *Ecology. 43*:410–432.

Raup, D. M. 1956. *Dendraster:* a problem in echinoid taxonomy. *Journal of Paleontology. 30*:685–694.

Raup, D. M. 1968. Theoretical morphology of echinoid growth. *Journal of Paleontology, Memoir. 2*:50–63.

Raup, D. M. Submitted. Taxonomic diversity during the Phanerozoic.

Raup, D. M., and Michelson, A. 1965. Theoretical morphology of the coiled shell. *Science. 147*:1294–1295.

Raup, D. M., and Seilacher, A. 1969. Fossil foraging behavior: computer simulation. *Science. 166*:994–995.

Raup, D. M., and Stanley, S. M. 1971. *Principles of Paleontology.* W. H. Freeman and Co. San Francisco. 388 pp.

Remane, A. 1952. *Die Grundlagen des natürlichen Systems, der vergleichenden Anatomie und der Phylogenetik.* Akademische Verlagsgesellschaft Geest & Portig. Leipzig. 400 pp. [Second edition, 1956.]

Reyment, R. A. 1958. Some factors in the distribution of fossil cephalopods. *Stockholm Contributions in Geology. 1*:97–184.

Reyment, R. A. 1969. Biometrical techniques in systematics. In *Systematic Biology, Proceedings of an International Conference conducted at the University of Michigan.* National Academy of Sciences, Publication No. 1692. Washington, D.C. pp. 542–587.

Reyment, R. A., and Van Valen, L. 1969. *Butonia olokundudui* sp. nov. (Ostracoda, Crustacea); a study of meristic variation in Paleocene and Recent ostracods. *Bulletin of the Geological Institution of the University of Uppsala.* New Series. *1*:83–94.

Rhoads, D. C. 1970. Mass properties, stability, and ecology of marine muds related to burrowing activity. In *Trace Fossils.* T. P. Crimes and J. C. Harper (Eds.). Seel House Press. Liverpool. pp. 391–406.

Rhoads, D. C., and Pannella, G. 1970. The use of molluscan shell growth patterns in ecology and paleoecology. *Lethaia. 3*:143–162.

Rhoads, D. C., and Young, D. K. 1970. The influence of deposit-feeding organisms on sediment stability and community trophic structure. *Journal of Marine Research. 28*:150–178.

Richter, R. 1929. Das Verhältnis von Function und Form bei den Deckelkorallen. *Senckenbergiana Lethaea. 11*:57–94.

Ricketts, E. F., and Calvin, J. 1952. *Between Pacific Tides.* Third edition. (Revised by J. W. Hedgpeth.) Stanford University Press. Stanford. 502 pp.

Ricklefs, R. E. 1970. Stage of taxon cycle and distribution of birds on Jamaica, Greater Antilles. *Evolution. 24*:475–477.

Riley, J. P., and Tongudai, M. 1967. The major cation/chlorinity ratios in sea water. *Chemical Geology. 2*:263–269.

Robie, R. A., and Waldbaum, D. R. 1968. Thermodynamic properties of minerals and related substances at 298.15°K (25°C) and one atmosphere (1.013 bars) pressure and at higher temperatures. *U.S. Geological Survey, Bulletin 1259.* 256 pp.

Robison, R. A., and Sprinkle, J. 1969. Ctenocystoidea: new class of primitive echinoderms. *Science. 166*:1512–1514.

Rogers, D. J., Fleming, H. S., and Estabrook, G. 1967. Use of computers in studies of taxonomy and evolution. *Evolutionary Biology. 1*:169–196.

Rollins, H. B., and Batten, R. L. 1968. A sinus-bearing monoplacophoran and its role in the classification of primitive molluscs. *Palaeontology. 11*:132–140.

Roughton, R. D. 1962. A review of literature on dendrochronology and age determination of woody plants. *State of Colorado, Department of Game and Fish, Technical Bulletin No. 15.* 99 pp.

Rowe, A. W. 1899. An analysis of the genus *Micraster*, as determined by rigid zonal collecting from the zone of *Rhynchonella Cuvieri* to that of *Micraster cor-anguinum. Quarterly Journal of the Geological Society of London. 55*:494–547.

Rudwick, M. J. S. 1961. The feeding mechanism of the Permian brachiopod *Prorichthofenia. Palaeontology. 3*:450–471.

Rudwick, M. J. S. 1962. Notes on the ecology of brachiopods in New Zealand. *Transactions of the Royal Society of New Zealand, Zoology. 25*:327–335.

Rudwick, M. J. S. 1964a. The function of zigzag deflections in the commissures of fossil brachiopods. *Palaeontology. 7*:135–171.

Rudwick, M. J. S. 1964b. The inference of function from structure in fossils. *The British Journal for the Philosophy of Science. 15*:27–40.

Runcorn, S. K. 1966. Corals as paleontological clocks. *Scientific American. 215*:26–33.

Russell, E. S. 1916. *Form and Function, a Contribution to the History of Animal Morphology.* J. Murray. London. 383 pp.
Ryther, J. H. 1969. Photosynthesis and fish production in the sea. *Science. 166*:72–76.

Sanders, H. L. 1968. Marine benthic diversity: a comparative study. *American Naturalist. 102*:243–282.
Sanders, H. L. 1969. Benthic marine diversity and the stability-time hypothesis. *Brookhaven Symposia in Biology. 22*:71–81.
Sanders, H. L., and Hessler, R. R. 1969. Ecology of the deep-sea benthos. *Science. 163*:1419–1424.
Sargent, M. C., and Austin, T. S. 1954. Biologic economy of coral reefs. Bikini and nearby atolls, Part 2. Oceanography (biologic). *U.S. Geological Survey, Professional Paper 260-E.* pp. 293–300.
Sattler, R. 1966. Towards a more adequate approach to comparative morphology. *Phytomorphology. 16*:417–429.
Schaeffer, B. 1965. The role of experimentation in the origin of higher levels of organization. *Systematic Zoology. 14*:318–336.
Schmidt, H. 1930. Über die Bewegungsweise der Schalencephalopoden. *Paläontologische Zeitschrift. 12*:194–207.
Schmidt, R. R., and Warme, J. E. 1969. Population characteristics of *Protothaca staminea* (Conrad) from Mugu Lagoon, California. *The Veliger. 12*:193–199.
Schoener, T. W. 1965. The evolution of bill size differences among sympatric congeneric species of birds. *Evolution. 19*:189–213.
Schoener, T. W. 1968. The *Anolis* lizards of Bimini: resource partitioning in a complex fauna. *Ecology. 49*:704–726.
Schopf, T. J. M. *In press.* Ergonomics of polymorphism: Its relation to the colony as the unit of natural selection. In *Symposium on Coloniality in Animals.* W. A. Oliver, Jr., R. S. Boardman, and A. H. Cheetham (Eds.). 81 ms pp.
Schopf, T. J. M., and Gooch, J. L. 1971. Gene frequencies in a marine ectoproct: a cline in natural populations related to sea temperature. *Evolution. 25*:286–289.
Schopf, T. J. M., and Manheim, F. T. 1967. Chemical Composition of Ectoprocta (Bryozoa). *Journal of Paleontology. 41*:1197–1225.
Seilacher, A. 1970. Arbeitskonzept zur Konstruktions-Morphologie. *Lethaia. 3*:393–396.
Seilacher, A. 1972. Divaricate patterns in pelecypod shells. (Konstruktions-Morphologie Nr. 4.) *Lethaia. 5*: 325–343.
Shaw, A. B. 1969. Adam and Eve, paleontology and the non-objective arts. *Journal of Paleontology. 43*:1085–1098.

Sheldon, R. W. 1965. Fossil communities with multi-modal size-frequency distributions. *Nature*. *206*:1336–1338.

Shiells, K. A. G. 1968. *Kochiproductus coronus* sp. nov. from the Scottish Viséan and a possible mechanical advantage of its flange structure. *Transactions of the Royal Society of Edinburgh*. *67*:477–507.

Shimer, H. W. 1914. *An Introduction to the Study of Fossils*. Macmillan. New York. 450 pp. [Second edition, 1929.]

Shrock, R. R., and Twenhofel, W. H. 1953. *Principles of Invertebrate Paleontology*. McGraw-Hill. New York. 816 pp.

Sibley, C. G. 1960. The electrophoretic patterns of avian egg-white proteins as taxonomic characters. *Ibis*. *102*:215–284.

Sibley, C. G. 1962. The comparative morphology of protein molecules as data for classification. *Systematic Zoology*. *11*:108–118.

Simberloff, D. S. 1969. Experimental zoogeography of islands. A model for insular colonization. *Ecology*. *50*:296–314.

Simberloff, D. S. 1970. Taxonomic diversity of island biotas. *Evolution*. *24*:23–47.

Simberloff, D. S., and Wilson, E. O. 1969. Experimental zoogeography of islands. The colonization of empty islands. *Ecology*. *50*:278–296.

Simberloff, D. S., and Wilson, E. O. 1970. Experimental zoogeography of islands. A two-year record of colonization. *Ecology*. *51*:934–937.

Simkiss, K. 1964. Phosphates as crystal poisons of calcification. *Biological Reviews*. *39*:487–505.

Simon, H. A. 1962. The architecture of complexity. *Proceedings of the American Philosophical Society*. *106*:467–482.

Simon, H. A. 1970. *The Sciences of the Artificial*. M.I.T. Press. Cambridge, Mass. 123 pp.

Simpson, G. G. 1944. *Tempo and Mode in Evolution*. Columbia University Press. New York. 237 pp.

Simpson, G. G. 1951. *Horses*. Oxford University Press. New York. 247 pp.

Simpson, G. G. 1953. *The Major Features of Evolution*. Columbia University Press. New York. 434 pp.

Simpson, G. G. 1961. *Principles of Animal Taxonomy*. Columbia University Press. New York. 247 pp.

Simpson, G. G. 1964. Species density of North American recent mammals. *Systematic Zoology*. *13*:57–73.

Slaughter, B. H. 1968. Animal ranges as a clue to late-Pleistocene extinction. In *Pleistocene Extinctions: the Search for a Cause*. P. S. Martin and E. H. Wright, Jr. (Eds.). Yale University Press. New Haven, Conn. pp. 155–167.

Slobodkin, L. B. 1961. *Growth and Regulation of Animal Populations*. Holt, Rinehart and Winston, Inc. New York. 184 pp.

Slobodkin, L. B., and Sanders, H. L. 1969. On the contribution of environmental predictability to species diversity. *Brookhaven Symposia in Biology*. *22*:82–93.

Smiley, C. J. 1969. Cretaceous floras of Changler-Colville Region, Alaska: stratigraphy and preliminary floristics. *American Association of Petroleum Geologists Bulletin*. *53*:482–502.

Sokal, R. R., and Sneath, P. H. 1963. *Principles of Numerical Taxonomy*. W. H. Freeman and Co. San Francisco. 359 pp.

Sokal, R. R., and Crovello, T. J. 1970. The biological species concept: a critical evaluation. *American Naturalist*. *104*:127–153.

Solomon, M. E. 1969. *Population Dynamics*. Institute of Biology, studies in biology no. 18. London. 60 pp.

Stahl, W. R. 1967. The role of models in theoretical biology. In *Progress in Theoretical Biology*. F. M. Snell (Ed.). Academic Press. New York. pp. 165–218.

Stanton, R. J., Jr., and Dodd, J. R. 1970. Paleoecologic techniques—comparison of faunal and geochemical analyses of Pliocene paleoenvironments, Kettleman Hills, California. *Journal of Paleontology*. *44*:1092–1121.

Stebbins, G. L. 1950. *Variation and Evolution in Plants*. Columbia University Press. New York. 643 pp.

Stehli, F. G. 1965. Paleontological technique for defining ancient ocean currents. *Science*. *148*:943–946.

Stehli, F. G. 1968. Taxonomic diversity gradients in pole location: the Recent model. In *Evolution and Environment*. E. T. Drake (Ed.). Yale University Press. New Haven, Conn. pp. 163–227.

Stehli, F. G., Douglas, R. G., and Newell, N. D. 1969. Generation and maintenance of gradients in taxonomic diversity. *Science*. *164*:947–949.

Stehli, F. G., and Wells, J. W. 1971. Diversity and age patterns in hermatypic corals. *Systematic Zoology*. *20*:115–126.

Stommel, H. 1963. Varieties of oceanographic experience. *Science*. *139*:572–576.

Straaten, L. M. J. U. van. 1960. Marine mollusc shell assemblages of the Rhône Delta. *Geologie en Mijnbouw*. *39*:105–129.

Swinnerton, H. H. 1923. *Outlines of Paleontology*. Edward Arnold & Co. London. 420 pp. [Section edition, 1930; third edition, 1949.]

Sylvester-Bradley, P. C. 1951. The subspecies in paleontology. *Geological Magazine*. *88*:88–102.

Sylvester-Bradley, P. C. (Ed.). 1956. *The Species Concept in Paleontology*. The Systematics Association Publication No. 2. Systematics Association. London. 145 pp.

Thomas, G. 1956. The species conflict. In *The Species Concept in Paleontology*. P. C. Sylvester-Bradley (Ed.). The Systematics Association Publication No. 2. Systematics Association. London. pp. 17–31.

Thompson, D'A. W. 1942. *On Growth and Form*. Second edition. Cambridge University Press. Cambridge, Mass. 1116 pp.

Tobias, P. V. 1965. Early man in East Africa. *Science*. *149*:22–33.

Tobien, R. 1949. Über die Lebenweise der Ascoceraten. *Neues Jahrbuch für Mineralogie, Geologie und Paläontologie, Abteilung B*. pp. 307–323.

Tobler, W. R., Mielke, H. W., and Detwyler, T. R. 1970. Geobotanical distance between New Zealand and neighboring islands. *Bio-Science*. *20*:537–542.

Trueman, A. E. 1922. The use of *Gryphaea* in the correlation of the Lower Lias. *Geological Magazine*. *59*:256–268.

Tullock, G. 1971. The coal tit as a careful shopper. *American Naturalist*. *105*:77–80.

Twenhofel, W. H., and Shrock, R. R. 1935. *Invertebrate Paleontology*. McGraw-Hill. New York. 511 pp.

Ubaghs, G. 1967. General characters of Echinodermate. In *Treatise on Invertebrate Paleontology*, Part S. R. C. Moore (Ed.). University of Kansas Press. Lawrence, Kansas. Echinodermata. *1*:S3–S60.

Urey, H. C., Lowenstam, H. A., Epstein, S., and McKinney, C. R. 1951. Measurement of paleotemperatures and temperatures of the Upper Cretaceous of England, Denmark, and the southeastern United States. *Geological Society of America Bulletin*. *62*:399–416.

Valentine, J. W. 1961. Paleoecological molluscan geography of the California Pleistocene. *University of California Publications of Geological Science*. *34*:309–442.

Valentine, J. W. 1968. The evolution of ecological units above the population level. *Journal of Paleontology*. *42*:253–267.

Valentine, J. W. 1969a. Niche diversity and niche size patterns in marine fossils. *Journal of Paleontology*. *43*:905–915.

Valentine, J. W. 1969b. Patterns of taxonomic and ecological structure of the shelf benthos during Phanerozoic time. *Palaeontology*. *12*:684–709.

Valentine, J. W. 1971a. Resource supply and species diversity patterns. *Lethaia*. *4*:51–61.

Valentine, J. W. 1971b. Plate tectonics and shallow marine diversity and endemism, an actualistic model. *Systematic Zoology*. *20*:253–264.

Valentine, J. W., and Moores, E. M. 1970. Plate-tectonic regulation of faunal diversity and sea level: a model. *Nature*. *228*:657–659.

Vernon, M. D. (Ed.). 1966. *Experiments in Visual Perception*. Penguin Books. Baltimore. 443 pp.

Vinogradov, A. P. 1953. The Elementary Chemical Composition of Marine Organisms. *Sears Foundation for Marine Research. Memoir 2.* 647 pp.

Vuilleumier, F. 1970. Insular biogeography in continental regions. 1. The northern Andes of South America. *American Naturalist. 104*:373–388.

Waddington, C. H. 1969. Paradigm for an evolutionary process. In *Towards a Theoretical Biology.* C. H. Waddington (Ed.). *2*:106–124.

Wallace, A. R. 1876. *The Geographical Distribution of Animals.* Harper & Bros. New York. Volume 1, 503 pp.; volume 2, 607 pp.

Wallace, A. R. 1889. *Darwinism: An Exposition of the Theory of Natural Selection With Some of Its Applications.* Macmillan. London. 494 pp.

Wallace, P., and Ager, D. V. 1966. Flume experiments to test the hydrodynamic properties of certain spiriferid brachiopods with reference to their supposed life orientation and mode of feeding. *Proceedings of the Geological Society of London. No. 1635.* pp. 160–163.

Walther, J. K. 1919–1927. *Allgemeine Paläeontologie; geologische Fragen in biologischer Betrachtung.* Gebrüder Borntraeger. Berlin. 809 pp.

Warme, J. E. 1969. Live and dead molluscs in a coastal lagoon. *Journal of Paleontology. 43*:141–150.

Watabe, N., and Wilber, K. M. 1960. Influence of the organic matrix on crystal types in molluscs. *Nature. 188*:334.

Watson, G. 1964. Ecology and evolution of passerine birds on the islands of the Aegean Sea. Ph.D. Thesis, Department of Biology, Yale University.

Webb, S. D. 1969. Extinction-origination equilibria in late Cenozoic land mammals of North America. *Evolution. 23*:688–702.

Weber, J. N. 1968. Fractionation of the stable isotopes of carbon and oxygen in calcareous marine invertebrates—The Asteroidea, Ophiuroidea and Crinoidea. *Geochimica et Cosmochimica Acta. 32*:33–70.

Weber, J. N., and Raup, D. M. 1966a. Fractionation of the stable isotopes of carbon and oxygen in marine calcareous organisms—the Echinoidea. Part I. Variation of C^{13} and O^{18} content within individuals. *Geochimica et Cosmochimica Acta. 30*:681–703.

Weber, J. N., and Raup, D. M. 1966b. Fractionation of the stable isotopes of carbon and oxygen in marine calcareous organisms—the Echinoidea. Part II. Environmental and genetic factors. *Geochimica et Cosmochimica Acta. 30*:705–736.

Weber, J. N., and Raup, D. M. 1968. Comparison of C^{13}/C^{12} and O^{18}/O^{16} in the skeletal calcite of Recent and fossil Echinoids. *Journal of Paleontology. 42*:37–50.

Weller, J. M. 1961. The species problem. *Journal of Paleontology. 35*:1181–1192.

Weller, J. M. 1969. *The Course of Evolution.* McGraw-Hill. New York. 696 pp.

Weymouth, F. W., and Rich, W. H. 1931. Latitude and relative growth in the razor clam, *Siliqua patula*. *Journal of Experimental Biology*. *8*:228–249.

White, M. J. D. 1954. *Animal Cytology and Evolution*. Second edition. Cambridge University Press. Cambridge, England. 454 pp.

Whitehead, D. R., and Jones, C. E. 1969. Small islands and the equilibrium theory of insular biogeography. *Evolution*. *23*:171–179.

Wilbur, K. M. 1964. Shell formation and regeneration. In *Physiology of Mollusca*. K. M. Wilbur and C. M. Yonge (Eds.). Academic Press. New York. pp. 243–282.

Wilbur, K. M., and Watabe, N. 1963. Experimental studies on calcification in molluscs and the alga *Coccolithus huxleyi*. *Annals of the New York Academy of Science*. *109*:82–112.

Williams, C. B. 1964. *Patterns in the Balance of Nature and Related Problems in Quantitative Ecology*. Academic Press. New York. 324 pp.

Willis, J. C. 1922. *Age and Area*. Cambridge University Press. Cambridge. 259 pp.

Wilson, E. O. 1959. Adaptive shift and dispersal in a tropical ant fauna. *Evolution*. *13*:122–144.

Wilson, E. O. 1961. The nature of the taxon cycle in the Melanesian ant fauna. *The American Naturalist*. *95*:169–193.

Wilson, E. O. 1968. The ergonomics of caste in the social insects. *The American Naturalist*. *102*:41–66.

Wilson, E. O. 1969. The species equilibrium. *Brookhaven Symposia in Biology*. *22*:38–47.

Wilson, E. O. 1971. *The Insect Societies*. Belknap Press, Harvard University. Cambridge. 548 pp.

Wilson, E. O., and Bossert, W. H. 1971. *A Primer of Population Biology*. Sinauer Associates, Inc. Publishers. Stamford, Connecticut. 192 pp.

Wilson, E. O., and Hunt, G. L. 1967. Ant fauna of Futuna and Wallis Islands, stepping stones to Polynesia. *Pacific Insects*. *9*:563–584.

Wilson, E. O., and Simberloff, D. S. 1969. Experimental zoogeography of islands. Defaunation and monitoring techniques. *Ecology*. *50*:267–278.

Wilson, E. O., and Taylor, R. W. 1967. An estimate of the potential evolutionary increase in species density in the Polynesian ant fauna. *Evolution*. *21*:1–10.

Wilson, J. B. 1967. Palaeoecological studies of shell-beds and associated sediments in the Solway Firth. *Scottish Journal of Geology*. *3*:327–371.

Winland, H. D. 1969. Stability of calcium carbonate polymorphs in warm, shallow sea water. *Journal of Sedimentary Petrology*. *39*:1579–1587.

Woods, H. 1893. *Elementary Paleontology: Invertebrate*. Cambridge University Press. Cambridge. 222 pp. [Eighth edition, 1946, 477 pp.]

Worthington, L. V. 1968. Genesis and evolution of water masses. *Meteorological Monographs*. *8*:63–67.

Wright, S. 1932. The roles of mutation, inbreeding, crossbreeding, and selection in evolution. In *Proceedings of the Sixth International Congress of Genetics. 1*:356–366.
Wright, S. 1967. Comments on the preliminary working papers of Eden and Waddington. In *Mathematical Challenges to the Neo-Darwinian Interpretation of Evolution.* P. S. Moorehead and M. M. Kaplan (Eds.). *The Wistar Institute Symposia, Monograph No. 5.* The Wistar Institute Press. Philadelphia. pp. 117–120.

Zangerl, R. 1948. The methods of comparative anatomy and its contribution to the study of evolution. *Evolution. 2*:351–374.
Ziegler, A. M. 1965. Silurian marine communities and their environmental significance. *Nature. 207*:270–272.
Ziegler, B. 1962. Die Ammoniten—Gattung *Aulacostephanus* im Oberjura (Taxionomie, Stratigraphie, Biologie). *Paläeontographica, A. 119*:1–173.
Zimmermann, W. 1959. Methoden der phylogenetic. In *Die Evolution der Organismen: Ergebnisse und Probleme der Abstammungslehre.* G. Herberer (ed.). Second edition. Fischer. Stuttgart. 2 volumes. pp. 25–102. [Also other editions.]
Zittel, K. A. von. 1895. *Grundzüge der Paläeontologie.* R. Oldenbourg Druck und Verlag. Munich. 971 pp.
Zittel, K. A. von. 1876–1893. *Handbuch der Paläontologie.* R. Oldenbourg Druck und Verlag. Munich and Leipzig. 5 volumes. 960 pp.
Zittel, K. A. von. 1896–1902. *Text-book of Paleontology.* Translated and edited by C. R. Eastman. Macmillan. New York. 2 volumes, in 3 parts. 706 pp. [Second edition, 1913–1925.]
Zittel, K. A. von. 1901. *History of Geology and Palaeontology to the end of the Nineteenth Century.* Walter Scott. Paternoster Square, London. 562 pp. [Translated by M. M. Ogilvie-Gordon. Reprinted by Hafner Publishing Co., New York. 1962.]

INDEX